PHim2
Prentice Hall Interactive Math

Key Concept Activity Lab Workbook

Intermediate Algebra

K. Elayn Martin-Gay
University of New Orleans

PEARSON
Prentice Hall

One Lake Street
Upper Saddle River, NJ 07458

Prentice Hall Interactive Math:
Project Manager: *Elaine Page*
Media Project Manager: *Audra Walsh*
Editor-in-Chief: *Christine B. Hoag*
Editor-in-Chief, Development: *Carol Trueheart*
National Program Consultant: *Diane Gray*
Executive Marketing Manager: *Eilish Collins Main*
Production Editor: *Donna Crilly*
Managing Editor, Production: *John Matthews*
Media Buyer: *Thomas Mangan*
Print Buyer: *Michael Bell*
Editorial Assistant: *Christina Simoneau*

Prentice Hall Media Group:
Director: *Dave Roh*
Executive Producer: *Philip Lanza*
Product Manager, Mathematics: *Ian Seekell*
Associate Product Manager: *Jennifer Mangan*
Senior Design Manager, Software and Print: *Greg Lamb*
Senior Production Manager: *Peter Silvia*
Senior Multimedia Developer: *Joshua Sutherland*
Multimedia Developers: *Chris Lynch, Robert McCarthy*
Voiceover Talent: *Bridget Page*
Production Engineers: *Christina Maestri, Eric Saulnier*
Software Developers: *Katherine Gregory, Tanya Allain, Lynne Stewart, Evan Mauser, Michael Duguay, Virginia Pyle, Brian Muller*

Pearson Prentice Hall
Pearson Education, Inc.
Upper Saddle River, New Jersey 07458

Printed in the United States of America

10 9 8 7 6 5 4 3 2 1

ISBN:0-13-183846-6

Pearson Education Ltd., *London*
Pearson Education Australia Pty. Ltd., *Sydney*
Pearson Education Singapore, Pte. Ltd.
Pearson Education North Asia Ltd., *Hong Kong*
Pearson Education Canada, Inc., *Toronto*
Pearson Educación de Mexico, S.A. de C.V.
Pearson Education—Japan, *Tokyo*
Pearson Education Malaysia, Pte. Ltd.
Pearson Education, *Upper Saddle River, New Jersey*

Table of Contents

Chapter 1—Real Numbers and Algebraic Expressions

1. Extension Exercise—Calculate Your BMI

2. Exploration Activity—Fill It Up

3. Conceptual Exercise—Fuel Efficiency

4. Group Activity—Singing in the Rain

5. Graphing Calculator Activity—Turning a Profit

1. Extension Exercise—Calculate Your BMI

For most people, Body Mass Index or BMI is a better indicator of an individual's risk of disease than body weight alone. To help diagnose a weight problem, health professionals use BMI, a measurement that gives a ratio between weight and height. Let's take a look at how BMI is measured and why it is important.

a. Use the following three-step procedure to calculate your BMI.

Step 1: Multiply weight (in pounds) by 703.

Step 2: Multiply height (in inches) by height (in inches) or find height squared.

Step 3: Divide the answer in Step 1 by the answer in Step 2 to get your BMI. Round to the nearest whole number.

BMI Category	**Health Risk Based Solely On BMI**
$BMI < 25$	Minimal
$25 \le BMI < 27$	Low
$27 \le BMI < 30$	Moderate
$30 \le BMI < 35$	High
$35 \le BMI < 40$	Very High
$BMI \ge 40$	Extremely High

See your physician for a thorough evaluation of your health risk.

b. Let the variable W represent weight in pounds and H represent height in inches. Write an algebraic expression for BMI based on the three-step procedure above.

BMI =

c. To check your new formula for BMI, recalculate your BMI using the formula from part *b*.

Calculation:

BMI score:

d. Now use the formula for BMI in part *b* to construct a table of BMI scores using the heights 5'6" and 5'11" and a range of possible weights from 120 pounds to 200 pounds in increments of 10. Round answers to the tenths place.

Weight, W (pounds)	$703W$	Height 5'6"		Height 5'11"	
		H^2	BMI	H^2	BMI
120					
130					
140					
150					
160					
170					
180					
190					
200					

e. Explain how you could construct a more extensive BMI table that might be used by health professionals to measure fitness and health risk for most of their patients.

____

____

____

f. When using the metric system, we can calculate BMI by taking the weight (in kilograms) divided by the height (in meters) squared,

$$\text{BMI} = \frac{W}{H^2}$$

We can convert weight from pounds to kilograms and height from inches to meters with the help of the following conversion factors:

1 meter ≈ 39.37 inches and 1 kilogram ≈ 2.2 pounds

Recalculate your BMI by substituting your weight in kilograms and your height in meters into the metric system formula above. Then see if your BMI score checks with your answers for parts *a* and *c*.

2. Exploration Activity—Fill It Up

People get their energy from calories in food while an automobile's energy comes from using gas as fuel. We can think of a car's mileage as how efficiently the car uses its fuel.

In the table below, we have five cars listed with their city mileages in miles per gallon. You will fill in this table as you complete the activity.

Car	Engine Size (liters)	City Mileage (mpg)	Fuel Capacity (gallons)	Distance on a tank (miles)	Cost of filling a tank ($)	Fuel Cost per mile ($)
Geo Metro	1.0	44	10.6			
Honda Civic DX	1.5	34	11.9			
Ford Taurus	3.0	19	16			
Lincoln Town Car	4.6	17	20			
Chevrolet Suburban	5.7	13	42			

(Source: *The New York Times*, May 12, 1996)

a. Based on the information given in the table, complete the following sentences. Support your answer with reasons.

In general, as the size of an engine increases, the city mileage ____________________

____

In general, as the size of an engine increases, the fuel capacity ____________________

____

b. Calculate the total distance that each car can travel on one tank of gas, assuming all gas is used. Round distance to the nearest mile and place your results in column five of the table.

c. What are the two factors that determine the distance a car can travel on a tank of gas?

First Factor: ____________________ Second Factor: ____________________

d. Out of the five cars listed, the Chevrolet Suburban and Geo Metro can travel the greatest distance on one tank of gas. For each car, write what factor has the greatest influence on creating more miles per tank of gas. Explain your reasoning.

Chevrolet Suburban: ______________________________

____

Geo Metro: ______________________________

____

e. Suppose the price of gasoline is $1.19 per gallon for regular and $1.37 per gallon for premium. If the Metro, Civic, and Taurus use regular and the Lincoln and Suburban use premium, then calculate the cost to fill an empty tank and place the result in column six of the table on the previous page. Round your answers to the nearest cent.

f. Calculate the cost of gas per mile for each vehicle and place the result in the last column of the table on the previous page. Round your answers to the nearest cent.

g. Make a general statement that compares the cost of driving the following pairs of cars.

Civic and Taurus: ______________________________

____

Civic and Suburban: ______________________________

____

h. Suppose you fill an empty tank in both the Geo Metro and the Honda Civic.

- How much further can you drive in the Geo Metro?
- How much money do you save filling an empty tank in the Geo Metro as compared with the Honda Civic?
- Recalculate fuel cost per mile by rounding to the thousandths place. Why is rounding the fuel cost per mile to the nearest cent in parts *e* and *f* misleading when comparing the cost of driving a Geo Metro to a Honda Civic?

3. Conceptual Exercise—Fuel Efficiency

In the Exploration Activity—Fill It Up, we looked at the cost of driving different types of cars. We can continue our investigation with a look at the dangers of round-off error. Assume that all cars use regular gas priced at $1.19 per gallon. Make sure to round all costs to the nearest cent.

Car	Fuel Cost per Mile	Weekly Cost	Yearly Cost	Monthly Cost
Geo Metro				
Honda Civic				

a. Calculate the fuel cost per mile for the Geo Metro and the Honda Civic as in the previous activity, but this time round your answer to the thousandths place and place your result in the second column of the table above.

b. In the third column, record the weekly cost assuming that you drive an average of 300 miles a week.

c. In the fourth column, record the yearly cost.

d. In the fifth column, record the monthly cost.

e. Find the difference in fuel cost per mile between the two cars. Write your answer as a decimal and as a fraction of one cent. Then describe the significance of the fraction in terms of the problem situation.

Decimal:

Fraction of one cent: ______________________________

f. What does this fraction of a cent savings in fuel cost per mile amount to over a week, a month, and a year (52 weeks)? Assume an average driving distance of 300 miles per week. Show all calculations and write each answer in a complete sentence.

Week: ______________________________

Month: ______________________________

Year: ______________________________

g. Suppose a large company leases 100 cars for its nationwide sales force. If the average salesperson travels 500 miles a week, then how much money will the company save on gas per year by leasing a Geo Metro instead of a Honda Civic? Assume each salesperson takes a two-week vacation and does not use the car during that period.

Calculations:

____

____

h. What other factors should the company consider when choosing which car to lease?

____

____

____

i. Add the Ford Taurus to the last row of the table on the previous page and fill in the different costs in each column. As before, assume you drive an average of 300 miles per week throughout the year and pay $1.19 per gallon of gas.

j. Find the difference in fuel cost per mile for the Geo Metro and the Ford Taurus. Write your answer as a decimal and in cents using a mixed number. Describe the significance of the mixed number in terms of the problem situation.

Decimal:

Mixed number in cents: ____________________________

____

k. What does the small savings in fuel cost per mile amount to over a week, a month, and a year (52 weeks)?

Week: __________________________________

Month: __________________________________

Year: __________________________________

4. Group Activity—Singing in the Rain

Seattle, Washington is often characterized as a rainy city, but is there heavy precipitation all year long? Let's find out by looking at some data. In the table below, assume that the month of January is number 1, the month of February is number 2, etc. and that precipitation is recorded in inches.

Monthly Normal Precipitation for Seattle, Washington

Month	1	2	3	4	5	6	7	8	9	10	11	12
Precipitation	5.4	4.0	3.8	2.5	1.8	1.6	0.9	1.2	1.9	3.3	5.7	6.0

(Source: National Climatic Data Center, U.S. Dept. of Commerce)

Construct a bar graph showing Seattle's normal monthly precipitation for each month of the year. January's bar has already been sketched as a model.

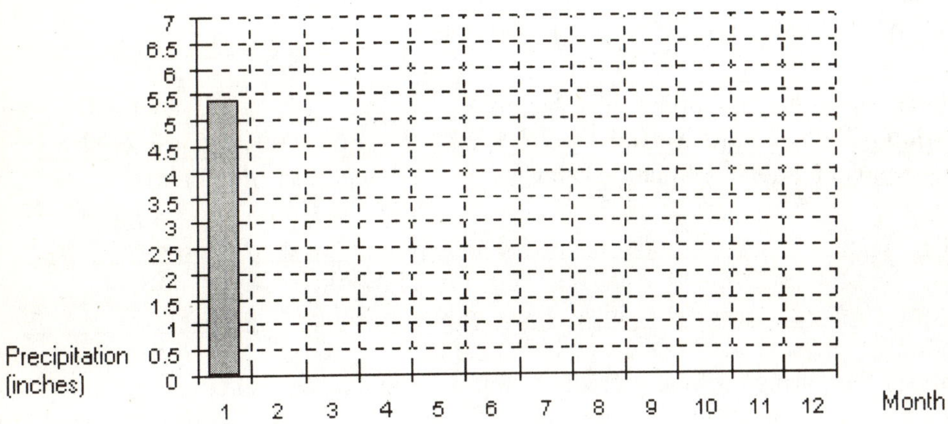

a. If you plan to visit Seattle in July, should you pack your umbrella? Explain your answer in a complete sentence.

____

____

____

b. In which months can you normally expect to receive more than 5 inches of precipitation?

____

c. During which two months did the precipitation increase the most? Explain how you can estimate your answer graphically and then verify it numerically using the table.

___

___

___

d. What was the decrease in precipitation from January to February? Show all work and write your answer in a complete sentence.

___

e. What was the normal amount of precipitation for the entire year in Seattle?

f. The monthly normal precipitation for Miami, Florida is given in the table below. What range of months would you define as the wet season? The dry season?

Month	1	2	3	4	5	6	7	8	9	10	11	12
Precipitation	2.0	2.1	2.4	2.9	6.2	9.3	5.7	7.6	7.6	5.6	2.7	1.8

(Source: National Climatic Data Center, U.S. Dept. of Commerce)

___

___

g. Write an analysis comparing the yearly precipitation patterns for the two cities.

___

___

___

___

___

___

___

5. Graphing Calculator Activity—Turning a Profit

Even though your math class has a demanding workload, you decide to start a simple landscaping business (mowing lawns, raking leaves and trimming bushes) to pay for next year's tuition. You advertise in the local paper for two weeks at a cost of \$128 and then buy the following yard tools: lawn mower (\$335), rake (\$16), and shrub clippers (\$21). The ad reads that you charge \$19 per hour.

a. What is the total cost to start your business?

b. The amount of money coming into your business is called the **revenue**. State the variable that revenue depends on and decide on an appropriate letter to represent your unknown.

c. Use your calculator as you complete the following table. Be sure to show the calculations needed to obtain the revenue for each amount of time worked (in hours). This will allow you to see the general pattern.

Time (hours)	Calculation	Revenue (\$)
10	$19 \cdot 10$	190
20		
30		
40		
50		
60		
70		
80		

d. Use the variable you chose from part *b* to write an algebraic expression for calculating revenue in terms of time.

e. Use your calculator as you complete another table that shows how to calculate profit as revenue minus cost.

Time (hours)	Calculation	Profit (\$)
10	$19 \cdot 10 - 500$	-310
20		
30		
40		
50		
60		
70		
80		

f. Write an algebraic expression for calculating profit in terms of time.

g. Before you mow any lawns, what is your company's profit?

h. How many hours must you work to make a positive profit? Answer this question by entering the algebraic expression for profit from part *f* into your calculator and using the table feature.

i. Set the algebraic expression for profit (from part *f*) equal to zero and solve for the variable representing time.

j. Explain the practical meaning of your answer to part *i* in terms of the problem situation.

___

___

___

k. Label the graph of each line below as the revenue, cost, or profit line.

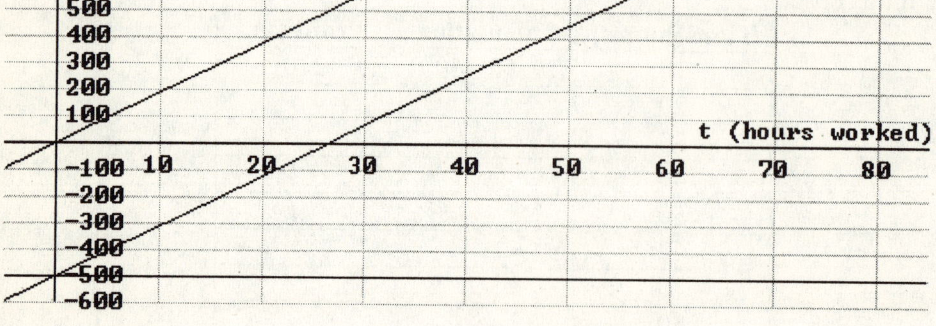

Chapter 2—Equations, Inequalities, and Problem Solving

1. Extension Exercise—Recreation Path

2. Exploration Activity—Back to School

3. Conceptual Exercise—Olympic Pole Vault

4. Group Activity—Buying a House

5. Graphing Calculator Activity—Growth of a Chain

1. Extension Exercise—Recreation Path

Let's investigate the relationship between distance, rate, and time by traveling to Stowe, Vermont. In Stowe, there is a recreation path where people walk, run, bike, and skate while enjoying the scenic beauty. The path starts at the center of town and meanders along a river and through a working dairy farm for 5 ¼ miles before it ends.

a. Suppose you are staying at a resort motel that is located near the path ¾ mile from where the path begins. Starting at the ¾ mile mark, you run until the 2 ¼ mile mark, then turn around and run back to where you began. What was the total distance of your run? Show all calculations.

b. If the run took 30 *minutes* to complete, what was your average speed or rate in miles per *hour*? *Hint:* $\text{average speed} = \dfrac{\text{distance traveled}}{\text{elapsed time}}$ or $r = \dfrac{d}{t}$

c. Suppose you had an electronic device that recorded the total elapsed time after each ¼ mile during the run. Using the formula from part *b,* record the average speed during each ¼ mile interval in the table below. *Hint:* Adapt the formula to be average speed equals change in distance over change in time.

Distance *d* (miles)	**Change in Distance**	**Time *t*** (minutes)	**Change in time** (minutes)	**Average Speed *r*** (miles/hour)
0		0		
¼		2 ½		
½		5		
¾		7 ½		
1		10 ½		
1 ¼		14 ½		
1 ½		17		
1 ¾		19 ½		
2		21		
2 ¼		23		
2 ½		25 ½		
2 ¾		28		
3		30		

d. How many miles had you run when you reached the turnaround point (2 ¼ mile mark)? Recall that you started your run at the ¾ mile mark. This corresponds to the first row in the table from part *c* or the point when total distance and time are zero.

e. Based on the different speeds calculated, write a possible description of what happened while running along the path.

____

____

____

____

f. The following day you rent a bike at the Mountain Bike Shop. On the same path, you ride from the ¾ mile mark to the 4 ½ mile mark. Then you turn around and travel back over the same route to where you started. How far did you ride during your 45-*minute* trip?

g. What was your average speed in miles per *hour* during the bike ride?

h. Upon returning to the Mountain Bike Shop, the manager hands you the store's policy on rental rates (see below). You hand him a 20% discount coupon that can be applied to the rental cost.

Rental Time	1 hr	1.5 hr	2 hr	2.5 hr	3 hr	3.5 hr	4 hr
Adult Bikes	$6	$9	$12	$14	$15	$15.50	$16

If you were gone with the bike for 2 hours and 40 minutes, then how much will you be charged for the rental after using the discount? *Hint:* Rental times that fall between intervals are charged the rental rate at the upper end of the interval. For example, a time of 1 hour and 35 minutes has the rental fee for 2 hours.

2. Exploration Activity—Back to School

Marissa is going back to school to get her teaching certificate. She needs 12 additional credits to obtain certification and plans to attend one of the two local colleges. City College costs $70 per credit plus a $150 one-time registration fee, while State College costs $95 per credit and has no registration fee. Assuming both colleges have a similar academic reputation, which is the better choice based on cost?

a. Show the computational steps necessary to find the cost of taking one 3-credit course at City College. Repeat the process for one 4-credit course at the same school.

b. Suppose the total number of credits a student takes is represented by the variable x and the cost to attend City College is represented by the variable C_1. Write a formula that expresses the cost C_1 in terms of the number of credits taken x.

c. Show the computational steps necessary to find the cost of taking one 3-credit course at State College. Repeat the process for one 4-credit course at the same school.

d. If the cost to attend State College is represent by C_2, then write a formula that expresses the cost C_2 in terms of the number of credits taken x.

e. Complete the table below to show the cost at each college for taking from one to twelve credits.

Number of Credits	Cost at City College C_1	Cost at State College C_2
1		
2		
3		
4		
5		
6		
7		
8		
9		
10		
11		
12		

f. Which college will cost Marissa the least amount of money to attend for 12 credits during one semester? What is that amount?

g. How much will Marissa pay if she decides to take 15 credits instead of 12 at the college in part *f*?

h. If it cost Marissa's friend Megan $1130 to attend City College last semester, how many credits did she take? Set up and solve the linear equation that will answer this question. Then verify your solution by extending the table you constructed in part *e*.

i. Use both inequality notation and interval notation to express solutions to the following questions.

- Over what range of credits does City College cost more than State College?
- Over what range of credits does City College cost less than State College?

j. When does City College cost exactly the same as State College?

____

____

k. Symbolically the cost at City College should be equal to the cost at State College when $C_1 = C_2$. Verify your answer to part *j* by setting the expressions from parts *b* and *d* equal to each other and then solving for *x*.

$$C_1 = C_2$$

3. Conceptual Exercise—Olympic Pole Vault

During the early years of the Olympics, the height of the winning pole vault increased as shown in the table below.

Time (year)	1900	1904	1908	1912
Height (inches)	130	138	146	154

a. Observe how the winning height increased over time from 1900 to 1912. Describe any pattern you see in the way time (input variable) changes with respect to height (output variable)?

____

____

____

b. According to this data, what would be the change in winning height per year?

c. Assuming the pattern in this data continues, write a verbal rule that explains the relationship between the independent variable, time in years, and the dependent variable, winning height in inches.

____

____

____

d. Suppose h is the height in inches and t is the number of years since 1900, so that 1900 would be time $t = 0$, 1904 would be time $t = 4$, etc. Write a symbolic rule in the form of an equation or formula.

$h =$ ____________________

e. Suppose the Olympics had been held in 1916 and the linear trend in the data table continued. Predict the winning height in 1916 by making the correct substitution into the equation developed in part *d.* Show all calculations that lead to your solution.

f. Again suppose the linear trend in the table continues over time. In what year would the Olympic pole vault champion jump 15 ½ feet? Set up and solve the proper equation. Then use complete sentences to write your solution in terms of the problem situation.

Calculations:

____

____

g. Read the actual Olympic records below and explain what you have learned about using a math model to predict data.

- U.S. athlete Frank Fross made the winning height of 13 feet 5 inches as recorded in the 1920 Olympics.
- U.S. athlete Don Bragg made the winning height of 15 feet 5 inches as recorded in the 1960 Olympics.

____

____

____

____

____

____

____

4. Group Activity—Buying a House

Kelly and Steven DuBois want to buy a house and need the minimum annual gross income necessary to get a mortgage. If the lowest interest rate available is 8%, then the following table shows us the income available for different loan amounts. The data assumes a 30-year loan and a monthly payment that cannot exceed 28% of the DuBois's gross income. Also, the data values do not include property taxes and insurance as part of the monthly payments.

Minimum Annual Gross Income	Loan amount
$15,724	$50,000
$23,586	$75,000
$31,447	$100,000
$47,171	$150,000
$62,895	$200,000

(Source: National Association of Realtors)

a. What are the two related variables (in words not symbols)?

b. Which variable depends on the other one?

c. Which variable is independent of the other one?

d. Describe the relationship between the two variables.

____

____

e. Suppose I represents the income needed to obtain a loan. Use an inequality symbol with the variable I to complete the following phrase.

A loan of $100,000 may be obtained if ______________________________

____.

f. If interest rates increase, how will the above table change to reflect that increase?

____

g. The above table assumes that the monthly payment cannot exceed 28% of the borrower's gross income. How would this table change if the 28% is lowered?

h. By how much does the Dubois's minimum annual gross income need to increase to make the following changes:

- A \$50,000 loan to a \$75,000 loan

- A \$75,000 loan to a \$100,000 loan

- A \$100,000 loan to a \$150,000 loan

- A \$150,000 loan to a \$200,000 loan

i. Use the table on the previous page to make a rough estimate of how much income is necessary to afford a home with a loan of \$90,000.

j. If Kelly and Steven's annual gross income is \$25,000, use the table to roughly estimate the largest loan they could obtain.

k. Think about how to obtain a nearly exact answer to parts *i* and *j*, and then complete any mathematical work necessary to arrive at that answer. Use complete sentences to explain the thought process behind your work.

5. Graphing Calculator Activity—Growth of a Chain

The following data shows how the opening of a single restaurant in 1983 grew into a successful chain of 198 stores as of May 1997. You will complete the last column later in this activity. For now, consider that time $t = 0$ is the year 1983, $t = 1$ is 1984, and so on.

Time (year) t	# of Stores n	Total Sales (millions of dollars)	Average Sales per Store (millions of dollars)
0	1	1.5	
1	2	3	
2	4	6	
3	8	16	
4	12	24	
5	20	38	
6	32	61	
7	47	90	
8	64	118	
9	93	160	
10	116	200	
11	137	250	
12	174	301	
13	189	325	

Source: (The Boston Globe, May 27, 1997)

a. Enter the given data above into three lists on a graphing calculator and graph a scatter plot of time t on the x-axis and number of stores n on the y-axis. Translate what you see on the screen to the coordinate system below.

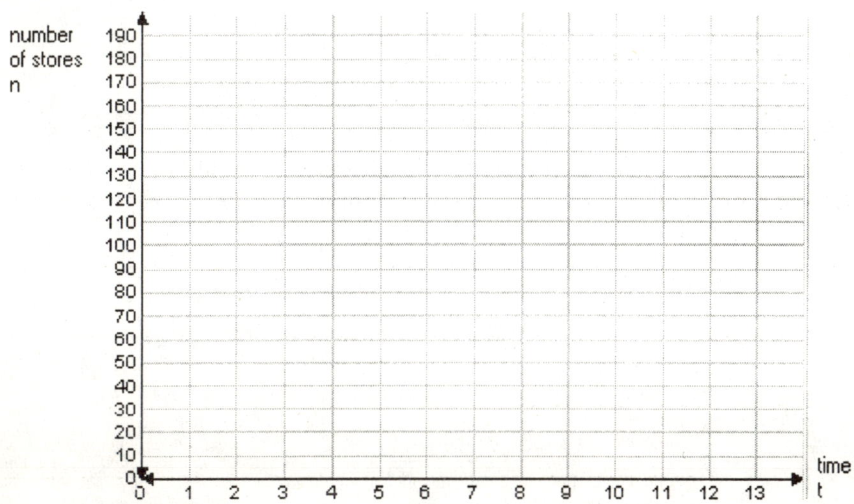

b. Use your graphing calculator to draw a new scatter plot of time versus total sales and explain how this plot compares with the sketch you made in part *a*.

c. Find a formula that calculates average sales per store for each year from 1983–1996. Use the words *number of stores* and *total sales* in developing the formula.

Average sales per store =

d. Create a fourth list in your calculator using the formula from part *c*. Enter the values into the fourth column of the table on the preceding page. On the lines below, explain any trend you see in the average annual sales per store over the given time period.

e. Suppose each franchise can serve about the same number of people. If the manager of one particular restaurant makes $1 million a year in total sales, how is his restaurant doing compared to the average franchise?

f. If the average bill is about $10 a person, then what was the average number of people served per day at each restaurant in 1996? Assume that the restaurant is open every day.

Chapter 3—Graphs and Functions

1. Extension Exercise—Alaskan Highs and Lows

2. Exploration Activity—Living on the Edge

3. Conceptual Exercise—Target Heart Rate

4. Group Activity—Trash Pickup

5. Graphing Calculator Activity—Global Warming

1. Extension Exercise—Alaskan Highs and Lows

Let's look at monthly temperature changes in Barrow, Alaska. Since temperature depends on the time of year, temperature is the dependent variable and appears along the vertical axis. Note that for each month (1-12) there is just one normal temperature in degrees Fahrenheit. The "normal" temperature for a specific month is the most frequently observed temperature for that month over time.

You can see this by starting at any month along the horizontal axis, follow the grid line vertically until you reach a point on the curve, and then move horizontally to the left until you reach the temperature on the vertical axis. The graph pictures temperature as a function of time because for each input (time) there is just one output (temperature).

Monthly Normal Temperatures for Barrow, Alaska

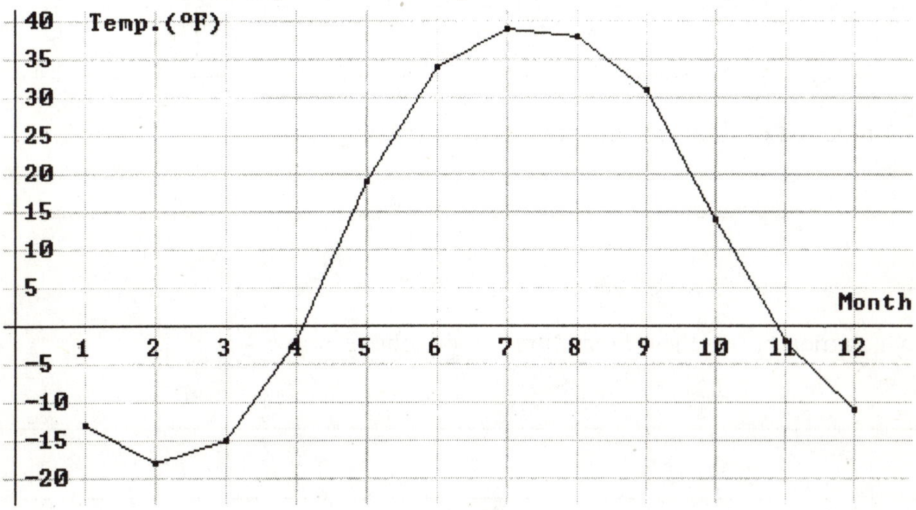

(Source: National Climatic Data Center)

a. Use the graph to estimate the coordinates for each month's data point and then fill in the table of ordered pairs. Let t be time (month) and T be temperature (°F). January's temperature has been recorded.

t	1											
T	-13											

b. If the temperature in Barrow is –2 °F, then it is most likely to be what month(s) of the year?

c. What month normally has the highest temperatures? What is that month's normal temperature? Write your answer in a complete sentence, in function notation, and as an ordered pair.

____

Function Notation: $T($___$)=$ Ordered Pair: (___, ___)

d. What month usually has the lowest temperatures? What is the normal temperature for that month? Write your answer in a complete sentence, in function notation, and as an ordered pair.

____

Function Notation: $T(\_\_\_)=$ Ordered Pair: $(\_\_\_,\_\_\_)$

Answer the following questions in complete sentences.

e. During what period of time is the temperature decreasing?

____

____

f. During what period of time is the temperature increasing?

____

____

g. During which month(s) is the temperature usually above zero?

____

____

h. During which two consecutive months did the temperature rise most rapidly?

____

____

2. Exploration Activity—Living on the Edge

The San Andreas Fault in California is a break in the earth's crust where many earthquakes have occurred. Parkfield is an area in central California that seismologists have studied to predict future earthquakes. The following data gives the years since 1857 in which the study center in Parkfield has measured major earthquakes at a magnitude of 6.0 on the Richter scale. Read the quake number as first, second, third, and so on.

Quake Number at 6.0 magnitude	1	2	3	4	5	6
Time (year)	1857	1880	1902	1925	1945	1967

a. Plot the ordered pairs of data as points on the rectangular coordinate system below. Then move a straight edge around until you obtain a best-fit line. This is a line that comes closest to all of the points but does not necessarily go through any of them. Draw your best-fit line.

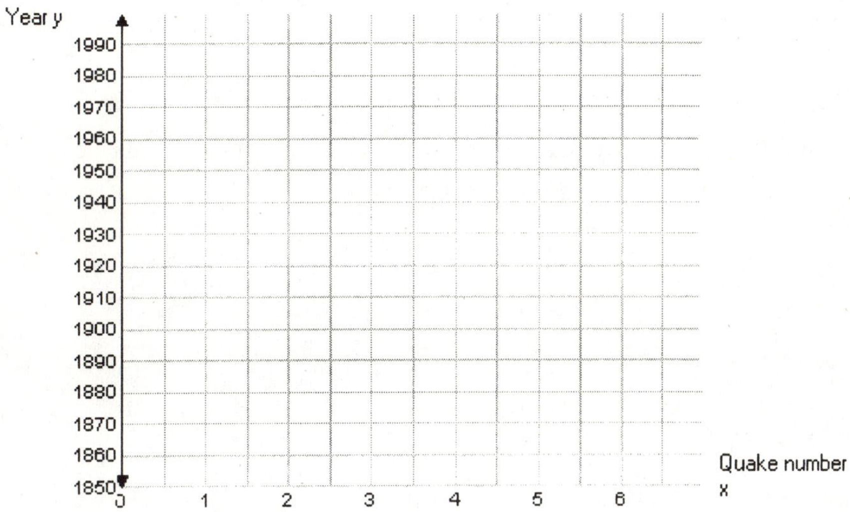

b. Find out how much time passed between earthquakes of magnitude 6.0 by finding each successive difference in y or change in time y (Δy).

Quake x (number)	**Time y (year)**	**Change in Time Δy**
1	1857	
2	1880	
3	1902	
4	1925	
5	1945	
6	1967	

What is the average amount of time between these earthquakes?

c. Use your graph from part *a* and your table from part *b* to predict when the seventh earthquake should have occurred. Explain your answer in complete sentences.

____

____

d. Estimate the coordinates of two points that fall on your best-fit line. Use these points to find the slope of the line. Show all work.

e. What does the value of your slope mean in terms of the problem situation?

____

____

f. Use the point-slope form of a linear equation to find the function $f(x)$ of your best-fit line.

$$y - y_1 = m(x - x_1)$$

$$f(x) =$$

g. Find $f(7)$ and explain what this means in the context of the Parkfield earthquake situation. How does your answer compare with results in part *c*?

$$f(7) =$$

____

____

3. Conceptual Exercise—Target Heart Rate

According to the American Heart Association, your target heart rate should stay within 50 to 75 percent of your maximum heart rate during exercise. The theory is to begin near 50% and work your way up to a maximum of 75% as you progress through a fitness program. The table displays the average maximum heart rate in beats per minute (bpm) that correspond to each age from 20 to 60 years in steps of 5 years.

Age x (years)	**Average Maximum Heart Rate M 100% (in bpm)**
20	200
25	195
30	190
35	185
40	180
45	175
50	170
55	165
60	160

(American Heart Association)

a. Let x represent age and M represent the average maximum heart rate. Explain why M is a function of x, and then give the domain and range of the function with respect to the table.

____

____

____

b. Describe the relationship between maximum heart rate and age by completing the sentence below.

As age increases by 5 years, average maximum heart rate __________________________

____

c. For each constant change in the input variable x, there is a constant change in output variable M. What type of function exhibits this type of behavior?

d. What is the slope of this function?

$$m = \frac{\text{Change in output}}{\text{Change in input}} = \frac{\Delta M}{\Delta x} =$$

e. Extend the beginning of the table on the previous page so that it includes 0, 5, 10, 15 as ages in the first column. Find the average maximum heart rate *M* for each of these ages.

Age *x*	**Average Maximum Heart Rate *M***
0	
5	
10	
15	

f. Write the equation that gives average maximum heart rate *M* as a function of age *x*. *Hint:* Use the slope-intercept form $f(x) = mx + b$ where the constant *b* is the output for the input 0 found in part *e*.

$$M(x) =$$

g. Suppose Steve, age 41, wants to start an exercise program. Use the rule from part *f* to find his maximum heart rate.

$$M(41) =$$

h. If Joe has an average maximum heart rate of 147 bpm, find his age by setting the function rule from part *f* equal to 147 and then solve for *x*.

i. Let's calculate Steve's target heart rate zone as 50-75%. Round heart rates to the nearest whole number.

- What is 50% of his average maximum heart rate?
- What is 75% of his average maximum heart rate?
- What is Steve's target heart rate zone using 50-75%?

j. Construct a table that gives the 50- 75% target heart rate zone for the ages listed. Round heart rates to the nearest whole number.

Age (Years)	**Target Heart Rate Zone 50 – 75% (in bpm)**
20	
25	
30	
35	
40	
45	
50	
55	
60	

4. Group Activity—Trash Pickup

In general, Americans are big on consumption. We love to buy things, use them, and ultimately throw them away. The table below contains data on how much garbage the United States has produced over time since 1960. Let time be represented as years since 1960, where 1960 is time $t = 0$, 1961 is time $t = 1$, etc.

Time t (year)	**Garbage g (millions of tons)**
0	90
10	120
20	150
30	180
40	210

(Source: EPA)

a. Based on the values shown in the table, is g a function of t? *Hint:* For each input t is there just one corresponding output g? If so, then state the domain and range. Otherwise explain why you think the relation is not a function.

____

____

b. How much garbage did the U.S. produce in 1960 ($t = 0$)? Answer the question in functional notation by using the table and completing the mathematical statement below.

$g(0) =$ ____________

c. Find the year since 1960 that the U.S. produced 150 million tons of garbage. Answer the question in function notation by using the table completing the mathematical statement below.

$g($____$) = 150$

d. Explain the meaning of the mathematical statement below in terms of the problem situation.

$g(30) = 180$

____

____

e. Explain the trend in the table by completing the following statement:

Every 10 years since 1960, ____.

f. Extend the table on the previous page to include columns for the change in time Δt, the change in garbage Δg, and the ratio of the change in garbage over the change in time $\frac{\Delta g}{\Delta t}$.

Hint: Change means to find the difference between successive values. Since $\frac{\Delta g}{\Delta t}$ is change in output over change in input, this ratio is also the slope of the function *g(t)*.

t	g	Δg	Δt	$\frac{\Delta g}{\Delta t}$
0	90			
10	120			
20	150			
30	180			
40	210			

g. Complete the following statement concerning slope:

If the amount of garbage increases 30 million tons every 10 years, then the amount of garbage increases ___________ million tons every year.

h. Plot each ordered pair given in the first two columns of the table in part *f* as a point in the rectangular coordinate system. Connect all the points with a straight line.

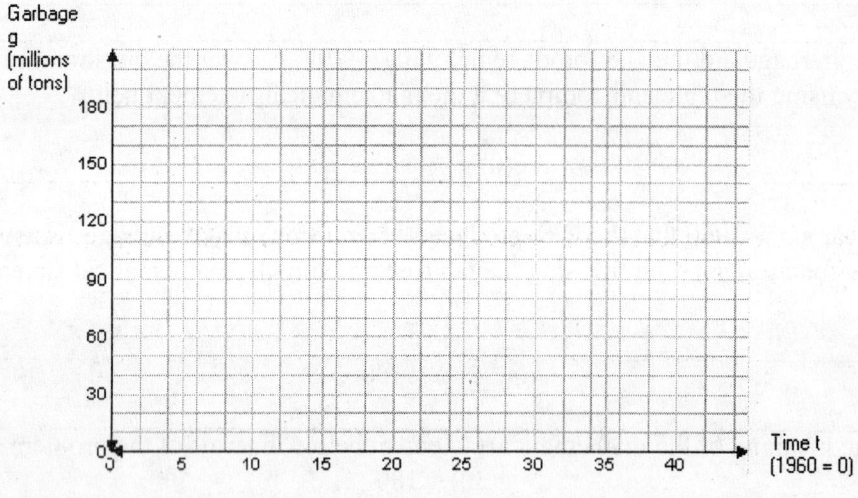

i. What type of function is graphed above? ______________________________

j. The line is a graphical representation of how the annual garbage produced in the U.S. has increased over time. Use a ruler to extend the line so that it reaches the end of the grid, then use the graph to predict how much garbage the U.S. will produce in the year 2010 (t = 50). Make sure that your answer follows the pattern in the table.

Prediction: ______________________________

k. What is the vertical intercept (y-intercept) for this function? ________________

l. Find an equation of the line that is a mathematical model of how much garbage is produced each year. Use the slope intercept form $g(t) = mt + b$ (generally $y = mx + b$) where m is the slope and b is the y-intercept.

$g(t) =$ ____________________

m. Use the model developed in part l to predict the amount of garbage produced in the U.S. in each of the following years: 1965, 1975, 1985, and 1995. Use function notation to express each prediction, show all calculations, and write a complete sentence that gives your answer in the context of the problem.

1965:

____

1975:

____

1985:

____

1995:

____

n. Do you have more confidence in your prediction for 1985 or 1995? Explain your answer using complete sentences.

____

____

____

5. Graphing Calculator Activity—Global Warming

About 80% of the yearly emission of carbon dioxide comes from burning fossil fuels such as coal, oil, and gas. Currently oceans and forests absorb only about 50% of carbon emissions caused from burning fossil fuels. The rest remains in the atmosphere. As carbon dioxide accumulates in the atmosphere, it traps the sun's heat. Some scientists believe that increasing concentrations of carbon dioxide and other greenhouse gases are causing global warming.

The table below gives the annual atmospheric concentrations of carbon dioxide for the period from 1965 to 1994.

Time (year)	Carbon Dioxide (parts per mill.)	Time (year)	Carbon Dioxide (parts per mill.)
1965	319.9	1980	338.5
1966	321.2	1981	339.8
1967	322.0	1982	341.0
1968	322.8	1983	342.6
1969	323.9	1984	344.3
1970	325.3	1985	345.7
1971	326.2	1986	347.0
1972	327.3	1987	348.8
1973	329.5	1988	351.3
1974	330.1	1989	352.7
1975	331.0	1990	354.0
1976	332.0	1991	355.5
1977	333.7	1992	356.2
1978	335.3	1993	357.0
1979	336.7	1994	358.8

(Source: Marland et. al., Oak Ridge National Laboratory, 1995)

a. Observe the change in carbon dioxide concentrations each year over the 30-year time period. Estimate the annual increase in carbon dioxide and describe any pattern you see in the data.

____

____

b. Enter the above data on a graphing calculator, letting 1965 = year 0, 1966 = year 1, 1967 = year 2, etc. Use the calculator to compute the rate of change in carbon dioxide concentration per year, called the annual increase. Then have the calculator sum each annual increase and divide the sum by 29. This gives the average annual rate of change in carbon dioxide concentration for the 30-year period.

c. Use your graphing calculator to plot the given data. Describe any pattern you see in the graph.

____

____

d. Develop a linear model of the form $f(x) = mx + b$ by using the average annual rate of change as the slope *m*. To find the *y*-intercept *b*, use the table to find output when input is zero. The input variable is time *t* and the output variable is carbon dioxide concentration *C*.

Slope $m =$ *y*-intercept $b =$

e. Use the slope and *y*-intercept to find a linear function $C(t)$ to model carbon dioxide concentration *C* as a function of time *t* in years since 1965.

Linear function $C(t) =$

f. Enter your linear function from part *e* into the graphing calculator and graph the line to see how close it comes to all of the data points. Does your model appear to be a good fit for the given data? Explain.

____

____

g. Use your linear function to predict the atmospheric concentration of carbon dioxide in the year 2010. Express your answer in function notation.

h. The 1994 ($t = 29$) carbon dioxide concentration is about 360 parts per million (ppm) in volume. Assuming the linear trend continues, when will carbon dioxide concentration double its 1994 level? Explain in complete sentences.

____

____

Chapter 4—Systems of Equations

1. Extension Exercise—Healthy Rabbits

2. Exploration Activity—Determining Dosage

3. Conceptual Exercise—Where Lightning Strikes

4. Group Activity—The Tortoise and the Hare

5. Graphing Calculator Activity—Consistent Systems?

1. Extension Exercise—Healthy Rabbits

Rabbits in a lab must be kept on a strict daily diet to include 30 grams of protein, 16 grams of fat and 24 grams of carbohydrates. A scientist has only three food mixes available with the following percentages of nutrients. Go through the problem-solving steps to find how many grams of each mix are needed daily to meet each rabbit's dietary needs.

	Protein	Fat	Carbohydrate
Mix A	31%	46%	23%
Mix B	67%	11%	22%
Mix C	23%	6%	71%

Understand the Problem:

a. Use the information above to determine the total grams of rabbit food needed daily to feed the rabbits.

Total grams of rabbit food: ________________

b. For each of the following combinations, verify that the total grams of rabbit food is the correct amount and determine the total grams of protein, fat, and carbohydrates in each mix and in the total.

	Total Grams of Each Mix	Grams of Protein	Grams of Fat	Grams of Carbohydrate
Mix A	35			
Mix B	22			
Mix C	13			
Totals (grams)				

	Total Grams of Each Mix	Grams of Protein	Grams of Fat	Grams of Carbohydrate
Mix A	8			
Mix B	34			
Mix C	28			
Totals (grams)				

c. ***Assign Letters and Illustrate:*** In the following chart, use the variables, a, b, and c, for the total number of grams of mixes A, B, and C, respectively. Show how the variables are used to fill in boxes of the table.

	Total Grams of Each Mix	**Grams of Protein**	**Grams of Fat**	**Grams of Carbohydrate**
Mix A				
Mix B				
Mix C				
Totals (grams)				

d. ***Translate the Problem into a System of Equations:*** Translate each of the following statements into an equation using the variables, a, b, and c.

"The final rabbit food mix has 30 grams of protein." ____________________

"The final rabbit food mix has 16 grams of fat." ____________________

"The rabbit food mix has 24 grams of carbohydrates." ____________________

e. ***Complete:*** Solve the system of equations from part d using determinants. Then use your graphing calculator to evaluate the values for each determinant. In the space below, show the determinants used and the results of your calculations. Round your final answer to one decimal place.

$a =$ $\frac{|\quad|}{|\quad|}$ $b =$ $\frac{|\quad|}{|\quad|}$ $c =$ $\frac{|\quad|}{|\quad|}$

$a =$ ________ $\approx$ ________ $b =$ ________ $\approx$ ________ $c =$ ________ $\approx$ ________

f. ***Interpret:*** On another piece of paper, write a note to the scientist explaining how many grams of each mix is needed to feed the rabbits the desired diet.

2. Exploration Activity—Determining Dosage

A 5% dextrose solution is sometimes used to feed a patient intravenously. It is also used as a base to dilute a drug that is to be administered intravenously. Suppose that Acme Pharmaceutical Products Inc., sells a 50 milliliter (ml) bottle that is 50% dextrose.

a. Determine the percent concentration of the total solution after adding different amounts of water to the 50-ml bottle of dextrose solution. Use the following table to record your results. If necessary, round to the nearest tenth of a percent.

Water Added (ml)	Total Solution (Water + 50 ml dextrose solution)	Percent Concentration* $\frac{25\ g\ dextrose}{total\ solution,\ ml} \times 100$
50		
100		
150		
200		
250		
300		
350		
400		
450		
500		

b. From observing the information above, how much sterile water would you add to the 50-ml bottle of 50% dextrose solution to get a 5% solution? How many milliliters of solution are in the final solution?

Water added: ___________ ml   Amount of 5% solution: ____________ ml

* When mixing a powder, such as dextrose, that is measured in grams, with a liquid measured in milliliters, we assume that 1 gram is approximately equal to 1 milliliter.

Now set up and solve a system of equations which illustrates the following problem:

You have a 50% dextrose solution and a 5% dextrose solution. How many milliliters of each need to be mixed in order to get 60 ml of a 30% dextrose solution?

c. Set up the system with a chart to organize data and assign variables:

Solutions	**Amount of Solution (in milliliters)**	**Dextrose Strength (in percent)**	**Amount of Pure Dextrose**
50% solution			
5% solution			
30% solution			

d. Translate the data into a system of two equations and solve. Round results to the nearest whole number.

e. A doctor has prescribed 499 milliliters of a 5% dextrose solution to be mixed with 1 milliliter of a Humulin R insulin. This should be administered to a patient intravenously at an infusion rate of 0.7 milliliters per minute. How long will it take to administer the 500 milliliters of the dextrose/insulin solution? Round results to the nearest whole. (Amount infused = infusion rate × time.)

time = ______________ minutes = ________________ hours

3. Conceptual Exercise—Where Lightning Strikes

Weather stations use a directional antenna to measure the electromagnetic field emitted by a lightning bolt. The antenna can determine the angle between a fixed point and the position of the lightning strike but cannot determine the distance to the lightning strike. However, the angle can be used to find the slope of the line connecting the fixed point (weather station) and the lightning strike. Then you can find the equation of this line. The coordinates of the lightning strike can be found by using data from two weather stations.

a. A weather station A is located at the coordinates (35 , 28), and a second station B is at (52 , 12). Plot the positions of the two weather stations on the grid below.

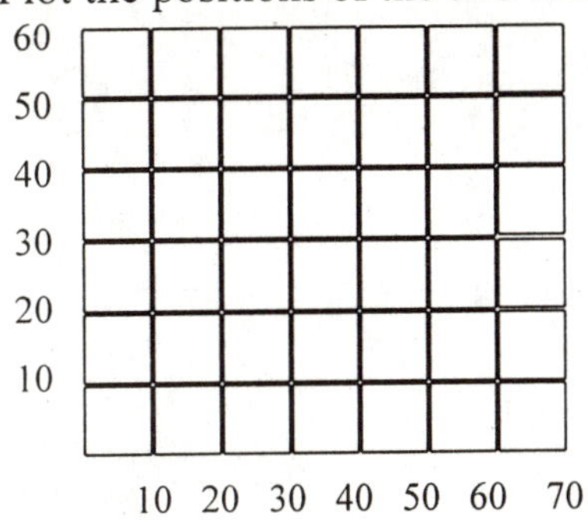

b. Suppose that a lightning strike is detected by both stations. The measured angle is used to calculate the slopes of the lines (on the grid) from the stations to the location of the lightning strike. For station A, the slope or m_a = -1.732, and for station B, m_b = 0.577. Write each of these ratios as a ratio of a number to 10.

$$m_a = \frac{-1.732}{1} = \frac{\quad}{10} \qquad\qquad m_b = \frac{0.577}{1} = \frac{\quad}{10}$$

c. Use the ratios above to sketch the line of vision (on the graph above) between each ranger station and the place where lightning struck. From the drawing, estimate the coordinates of the point where the lightning struck.

Point (,   )

d. Using the coordinates of station A and the slope m_a, find the equation of the line connecting station A and the lightning strike position.

e. Repeat the process in part *d* for the line connecting Station B and the point where lightning struck.

f. The two equations from parts *d* and *e* form a system of equations. Solve the system using either substitution or elimination.

g. Solve the system of equations using matrices.

h. Use a graphing calculator to solve the system of equations by graphing.

i. How does the solution from part *h* compare with your estimate in part *c* ? Explain in words how the solution relates to the problem situation.

____

____

____

4. Group Activity—The Tortoise and the Hare

In the classic story, "The Tortoise and the Hare," the tortoise and the hare engage in a race. The hare, knowing that he is much faster than the tortoise, takes a long break and loses the race because he naps too long. In this group activity, one student will be the tortoise and walk a distance while the other student will be the hare and run the same distance.

a. In the hallway or on a sidewalk, mark "Start" and "Finish" lines thirty feet apart. One student, the tortoise, will walk the distance while another group member clocks the time (in seconds). Another student, the hare, will run the length of the path while the other group member clocks the time.

Distance __________ Tortoise's time __________ Hare's time __________

b. Use the information above and the equation, $d = rt$, to compute the rates r (in feet per second) of the tortoise and the hare. Then substitute the calculated values for r into the equation $d = rt$ to get equations that describe the distance the tortoise and hare travel (on average) in t seconds.

Tortoise's rate: $r =$ __________ Hare's rate: $r =$ __________

Equation for tortoise: $d =$ ______ $> t$ Equation for hare: $d =$ ______ $> t$

c. Plot the graphs of the two equations from part b on the grid shown below. Pick an appropriate scale for the t (seconds) axis that shows what would happen in a 30-foot race.

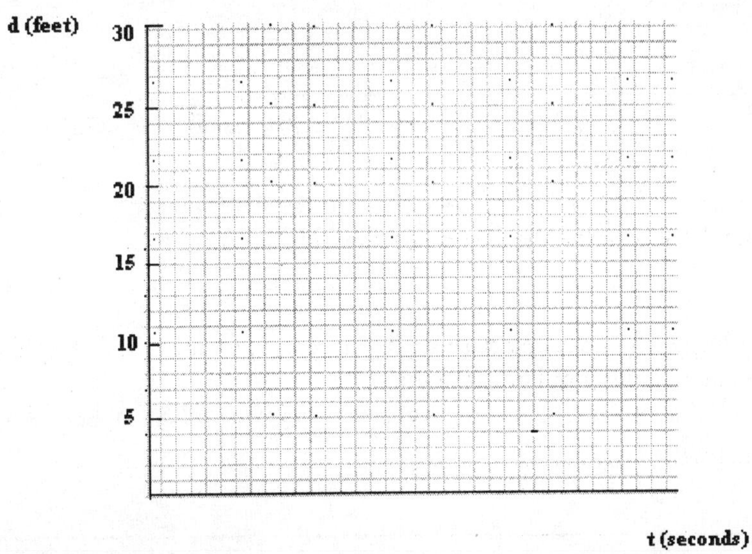

d. What is the slope of each line and what is its significance to the race?

Tortoise's slope: ________________ Hare's slope: ________________

____

____

e. If the tortoise is given a 10-foot head start in a race, write equations that represent the distance the tortoise and the hare will be from the starting line after t seconds.

Tortoise: ________________ Hare: ________________

f. Form a system of equations from the two equations in part *e*, and solve the system using the substitution method. Sketch the system and the solution on the grid below. Choose an appropriate scale for the t (seconds) axis.

System of equations and solution:

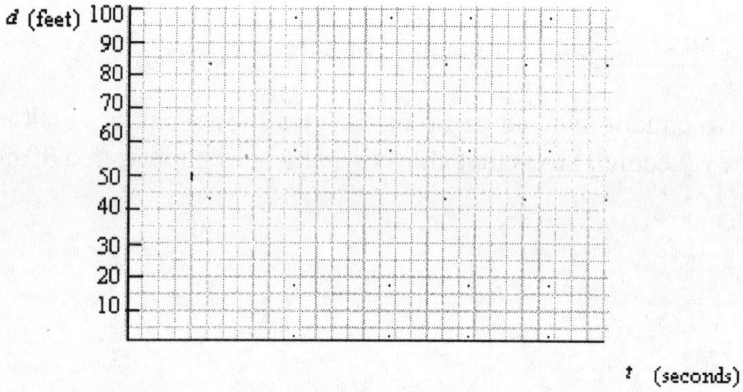

g. Explain the significance of the point of intersection in terms of the problem situation.

____

____

h. Predict who would win a 50-foot race if the tortoise was given a 20-foot head start? Justify your answer mathematically.

____

____

5. Graphing Calculator Activity—Consistent Systems?

Using your graphing calculator to graph each system of equations in parts *a-c*. Sketch your graphs on this worksheet and then determine whether each system is consistent with a single solution, dependent, or inconsistent. If the system has a single solution, record the *x*- and *y*-values correct to two decimal places.

a. Solve each equation in the system for *y*.

$17x - 5y = 215$ $y =$ ____________________

$-22x - 32y = -351$ $y =$ ____________________

On your graphing calculator, graph the two lines in the system. Choose a window that shows the graphs. If the system has a single solution, then choose a window that shows the solution point. Show the graph of the lines below, and label the maximum and minimum *x*- and *y*-values for the graphing window you chose.

Graph:

y max: _____

y min: _____

x min: _____ *x* max: _____

If there is a single solution, write it below. Round results to two decimal places.

$x =$ ________

$y =$ ________ solution as ordered pair: (,)

Is the system consistent, dependent, or inconsistent? Explain how you know.

____

____

____

b. Solve each equation in the system for *y*.

$22x + 9y = 1000$ $y =$ ____________________

$-66x - 27y = -50$ $y =$ ____________________

On your graphing calculator, graph the two lines in the system. Choose a window that shows the graphs. If the system has a single solution, then choose a window that shows the solution point. Show the graph of the lines below, and label the maximum and minimum *x*- and *y*-values for the graphing window you chose.

Graph:

y max: _____

y min: _____

x min: _____ *x* max: _____

If there is a single solution, write it below. Round results to two decimal places.

$x =$ ________

$y =$ ________ solution as ordered pair: (,)

Is the system consistent, dependent, or inconsistent? Explain how you know.

____

____

____

c. Solve each equation in the system for y.

$22x + 9y = 1000$ $y =$ ____________________

$-66x - 27y = -3000$ $y =$ ____________________

On your graphing calculator, graph the two lines in the system. Choose a window that shows the graphs. If the system has a single solution, then choose a window that shows the solution point. Show the graph of the lines below, and label the maximum and minimum x- and y-values for the graphing window you chose.

Graph:

y max: ____

y min: _____

x min: _____ x max: _____

If there is a single solution, write it below. Round results to two decimal places.

$x =$ ________

$y =$ ________ solution as ordered pair: (,)

Is the system consistent, dependent, or inconsistent? Explain how you know.

____

____

____

d. From observing the systems of equations in parts *a, b,* and *c*, explain how you might determine whether the system is consistent, inconsistent, or dependent from the equations and graphs of the lines.

Chapter 5—Exponents, Polynomials, and Polynomial Functions

1. Extension Exercise—Airline Bag

2. Exploration Activity—Wind Power

3. Conceptual Exercise—Pythagorean Puzzle

4. Group Activity—Water Density

5. Graphing Calculator Activity—Identifying Sign Errors

1. Extension Exercise—Airline Bag

A United Airlines sign in San Francisco states that passengers can carry at most two bags on the plane. The total of the dimensions ($L + W + H$) of each bag must not exceed 45 inches. Courtney wants to find a carry-on bag of maximum volume that will hold her tripod which collapses to 18 inches long. She decides to allow an extra inch in the bag so that the tripod will easily fit.

a. If the length, L, of the bag is 19 inches, how long is the combined width and height of the bag (all three dimensions add to 45 inches)?

$W + H =$ ______________

b. Solve the equation from part *a* for W.

c. Substitute the value for L and the expression for W (from part *b*) into the formula for volume, $V = LWH$, to express V as a function of the variable H. (In computing the volume, we are ignoring the width of the material. The actual volume will be slightly less because of the space the material itself uses.)

$V(H) =$ ______________

d. Write the expression from part *c* in another form by multiplying the factors.

$V(H) =$ ______________

e. Compute $V(H)$ for each of the values of H below. Record your answers in the table.

Value for *H* **(inches)**	***V(H)*** **(cubic inches)**
6	
9	
12	
15	
18	

f. What height from part *e* gives the greatest volume for the bag? What is the volume of the bag with that height? Include correct units with your answers.

height $H =$ __________ volume $V(H) =$ __________

g. Pick an appropriate window and graph *V(H)* on your graphing calculator. Find the high point or maximum of the graph. Sketch the graph of *V(H)* on the grid below, labeling the coordinates of the high point. State in words what each coordinate represents in terms of the bag. Include correct units with your explanation.

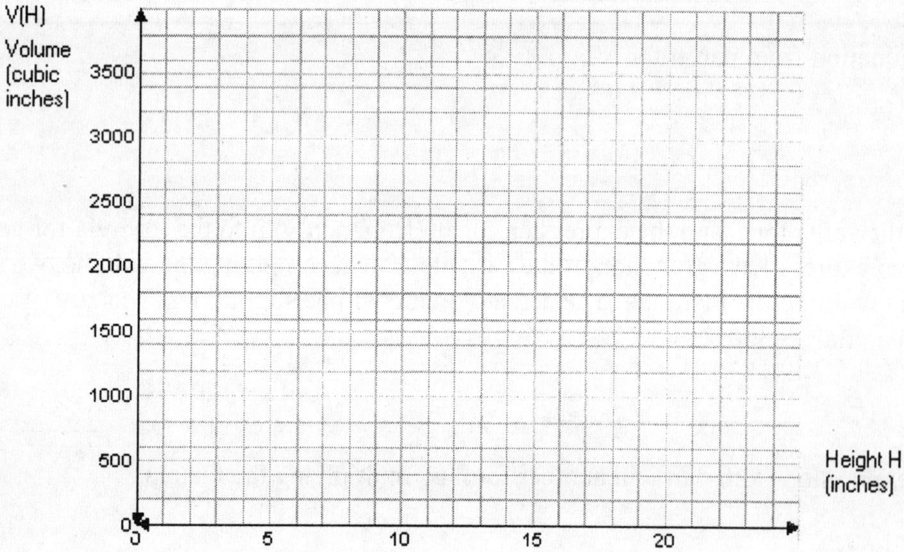

High point: (,) __________

____

h. What are the dimensions of the bag of greatest volume that fall within the required specifications? Include correct units with your answers.

$L =$ __________ $W =$ __________ $H =$ __________

2. Exploration Activity—Wind Power

One way to generate electricity is to use wind power. For moderate wind speeds, a windmill will generate watts of electrical power according to the following formula.

$P(v) = 0.015v^3$ P is the power, measured in watts; v is wind speed measured in miles per hour

a. Use the function, $P(v)$, to determine how much power is generated by the following wind speeds. In the table, show the values substituted into the formula, then the results of the computations. In the third column, round $P(v)$ to the nearest whole number.

Wind Speed--*v* (miles per hour)	***P(v)*** Show Substitutions	**Power, *P(v)*** (watts)
5		
10		
15		
20		
25		

b. On the grid below, graph $P(v)$ for wind speeds from 0 to 25 miles per hour.

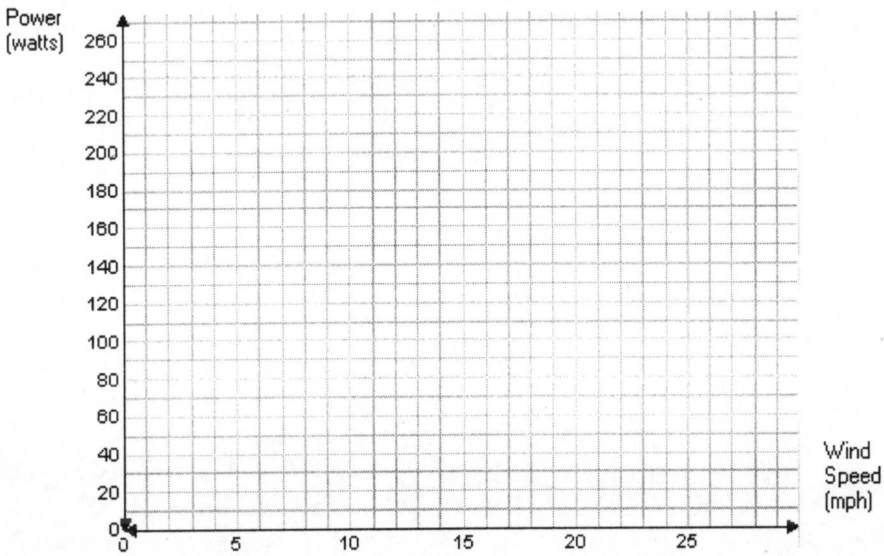

c. Determine the wind speed v necessary to create enough power to light each of the following light bulbs. Use the graph from part *b* to estimate the wind speed in miles per hour.

Type of Light Bulb	**Estimated Wind Speed** (miles per hour)
40-watt bulb	
60-watt bulb	
75-watt bulb	
100-watt bulb	

d. Suppose the Kendall family lives on the windy plains where the average wind speed is about 19 miles per hour. How much power does one windmill generate on average? Show or explain how you arrived at your answer.

e. How many windmills do the Kendalls need to generate one kilowatt (1000 watts) of energy? Show or explain how you arrived at your answer. Round to the nearest whole number.

3. Conceptual Exercise—Pythagorean Puzzle

a. On the grid below, Triangles 1 and 2 are right triangles. Count the units on the grid to verify that these two triangles are congruent. Use the letter a to label the two shorter legs of Triangle 1 and Triangle 2, and label the longer legs b. Use the letter c to label the hypotenuse of Triangles 1 and 2.

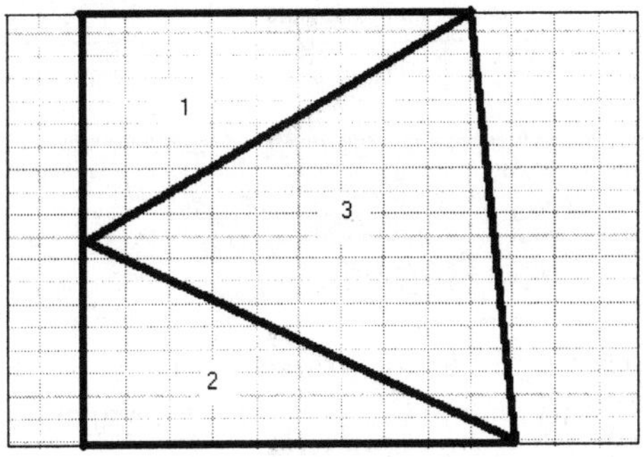

Triangle 1	**Triangle 2**
$a =$ ______	$a =$ ______
$b =$ ______	$b =$ ______
$c =$ ______	$c =$ ______

b. In terms of the variables a, b, and c, write expressions for the areas of Triangles 1 and 2. Find and simplify the sum of the two areas. *Hint:* $A = \frac{1}{2}bh$. The base and height of a triangle form right angles.

Triangle 1: $A =$ ______________ Triangle 2: $A =$ ______________

Sum: $A =$ ______________ = ______________
simplified

c. When congruent triangles are placed similarly as Triangles 1 and 2, the hypotenuse of one triangle is perpendicular to the hypotenuse of the other. In terms of c, what is the area of Triangle 3 on the drawing?

Triangle 3: $A =$ ______________

d. Assume one square on the grid is one unit. Compute the slopes of the lines defined by the hypotenuses of Triangles 1 and 2. Use your results to explain why the hypotenuse of one triangle is perpendicular to the hypotenuse of the other triangle. *Hint:* Count squares to determine the rise, or drop, and run of each line.

Slope 1: ______________________ Slope 2: ______________________

____

____

____

e. The outside lines of the entire drawing form a trapezoid. In terms of *a, b,* and *c,* find the area of the trapezoid by adding the expressions found in parts *b* and *c*. Simplify your answer.

f. Use the formula, $A = \frac{1}{2}(b_1 + b_2)h$, to find the area of the trapezoid in terms of *a* and *b*.

Hint: The bases, b_1 and b_2, are the two parallel sides and the height, *h*, is perpendicular to the two bases.

g. Explain why the two expressions from parts *e* and *f* are equal.

____

____

____

____

____

h. Set the two expressions from parts *e* and f equal to each other and solve the resulting equation for c^2 to get the Pythagorean theorem that relates the two legs of a right triangle to the hypotenuse. If your final result is not the Pythagorean theorem, check your work in this step or in previous steps.

Congratulations! You just reconstructed a proof of the Pythagorean theorem that our twentieth president, James Garfield, discovered.

The Pythagorean theorem is: $a^2 + b^2 = c^2$, where a and b are the legs of a right triangle and c is the hypotenuse. If you solve the Pythagorean theorem for c, you get the following result.

$$c = \sqrt{a^2 + b^2}$$

Illustrate each of the following situations on the grids below. Then determine the distance that the two people are apart. Measure the distance on the grid and calculate the distance. In the illustration, label each distance. *Hint:* When doing calculations on a calculator, be sure to insert parentheses around the entire expression under the square root symbol.

i. Kim and Lance start walking from the same point. Kim walks 11 feet south, and Lance walks 9 feet west. Let one square on the grid represent 1 square foot.

Measured distance apart: ____________

$c \approx$ ____________________ miles
measured

Calculated distance apart: __________

$c \approx$ ____________________ miles
calculated

j. Cheryl and Dan start driving from the same place. Cheryl drives 22 kilometers east, and Dan drives 17 kilometers north of the starting point. Let one square on the grid represent 1 square kilometer.

Measured distance apart: ____________

$c \approx$ ____________________ miles
measured

Calculated distance apart: __________

$c \approx$ ____________________ miles
calculated

4. Group Activity— Water Density

Density, a measure of the compactness of a material, is commonly measured in pounds per cubic foot, tons per cubic foot, or grams per cubic centimeter. In this activity, each group will work with solids made of three different materials, estimate the volume of each solid by the amount of water it displaces, weigh each solid, and then use a formula to compute the solid's density. The groups will then use the formula in another context where volume and mass are given in scientific notation.

Formula for density: $D = \frac{M}{V}$ D is density, M is mass (or weight), V is volume

The units must match when using the formula. For example, if mass is measured in pounds, and volume in cubic inches, then density is measured in pounds per cubic inch.

Materials needed for each group:

- 3 small solids of different densities--such as a rock, a rubber ball, a piece of Styrofoam
- A coffee tin or other cylindrical canister that will hold water
- A ruler (preferably a thin one)
- A scale that measures pounds (or ounces)

a. In the space below, describe your three solids. Which solid seems to be most dense (or with particles compact and pressed closest together), and which solid seems to be least dense? Explain how you arrived at your conclusion.

____

____

____

b. Determine the mass of each solid object by finding its weight in pounds. *Hint:* 16 ounces = 1 pound

$M_1 =$ ____________ lb $M_2 =$ ____________ lb $M_3 =$ ____________ lb

Put water in the canister to about ¾ full, and check which solids float and which solids sink.

c. If a solid has density less than that of water, it floats. Which of your solids appears to have density less than that of water? Which solid(s) appear(s) to have density greater than water?

Solids with less density than water: ____________ Solids with greater density than water: ____________

d. One at a time, place each solid in the canister and submerge the solid in water. Hold the solid under water if necessary. To the nearest $\frac{1}{16}$ inch expressed in decimal form, measure the height of the water and the diameter of the container. Compute the volume of water and solid together using the formula, $V = \pi r^2 h$. *Hint:* The radius, r, is half of the diameter.

Remove the solid from the water and find the volume of the water. Compute the volume of the solid by subtracting the volume of the water from the combined volume of the solid and water. Round your volumes to the nearest 0.5 cubic inch, and record your results in the table below. Include correct units with your answers.

Solid (name)	**Water + Solid**	**Water**	**Volume of Solid** (Col 2 – Col 3)
	$h =$ $r =$ $V =$	$h =$ $r =$ $V =$	$V =$
	$h =$ $r =$ $V =$	$h =$ $r =$ $V =$	$V =$
	$h =$ $r =$ $V =$	$h =$ $r =$ $V =$	$V =$

e. Using the formula for density, compute the density of each solid.

Solid (name)	**Volume, *V*, of Solid** (cubic in.)	**Weight, *W*, of Solid** (pounds)	**Density, $D = M/V$** (pounds per cu. in.)

f. The density of ordinary water is about 3.2×10^{-2} tons per cubic foot. Write the density of each solid in scientific notation. Convert the density of water to pounds per cubic inch, and compare the density of water to the density of each solid. Do your comparisons confirm your observations in part *c*? Explain your answers and conclusions. *Hint:* 1 ton = 2000 pounds and 1 cubic foot = 12^3 cubic inches

5. Graphing Calculator Activity—Identifying Sign Errors

a. All of the following factored expressions are incorrect factorizations of $6x^2 - 5x - 1$. Figure out what is wrong with each factorization by multiplying the two factors in each case. Which option can be corrected easily with a simple sign change? Out of the three choices below, which expression most closely matches the graph of $y = 6x^2 - 5x - 1$?

$(6x - 1)(x + 1)$ $(3x - 1)(2x + 1)$ $(1.01x - 1)(5.99x + 1)$

____

____

b. On your graphing calculator, assign $y_1 = 6x^2 - 5x - 1$, $y_2 = (6x - 1)(x + 1)$, $y_3 = (3x - 1)(2x + 1)$ and $y_4 = (1.01x - 1)(5.99x + 1)$. One at a time, compare the graphs of y_2, y_3, and y_4 to the graph of y_1. Which of the graphs most closely resembles y_1?

____

c. In the table below, calculate $6x^2 - 5x - 1$ and the three factored expressions for the values of x given in the first column. To do the calculations quickly, store the value into x, then call up the appropriate function to do the calculations.

$x =$	$6x^2 - 5x - 1$	$(6x - 1)(x + 1)$	$(3x - 1)(2x + 1)$	$(1.01x - 1)(5.99x + 1)$
3				
-5				
4				
7				
-10				

d. Compare the columns in the table. Which factorization appears to calculate approximately the same as $6x^2 - 5x - 1$?

____

____

When an expression is factored correctly, the equation formed by setting the original expression equal to its factored form is called an ***identity.*** This means that the same output occurs when the same number is substituted into both expressions.

e. Factor $6x^2 - 5x - 1$. Check your answer by comparing the graphs of $y_1 = 6x^2 - 5x - 1$ and $y_2 =$ the factored form.

$6x^2 - 5x - 1 =$ ____________________

For each of the following:

- Factor the expression. Check your work by graphing y_1 and y_2.
- Use the "trace" or "calc" feature on your graphing calculator to find all x-intercept points.
- Solve the related equation.

f. Factor: $8x^3 + 14x^2 - 30x =$ ____________ Solve: $8x^3 + 14x^2 - 30x = 0$

Graph: $y_1 = 8x^3 + 14x^2 - 30x$

$y_2 =$ ______________ (factored form) Solutions: $x =$ ____________

x-intercepts: ________________

g. Factor: $9x^2 + 6x + 1 =$ ____________ Solve: $9x^2 + 6x + 1 = 0$

Graph: $y_1 = 9x^2 + 6x + 1$

$y_2 =$ ______________ (factored form) Solutions: $x =$ ____________

x-intercepts: ________________

h. Use parts f and g to explain how the x-intercepts of a polynomial $p(x)$ are related to the solutions of $p(x) = 0$.

____

____

____

____

i. If you are asked to factor $4x^2 - 9$, and the result is $(2x - 3)(2x + 3)$ with a final answer of $x = 3/2$ or $x = -3/2$, what is wrong with the answer?

Chapter 6—Rational Expressions

1. Extension Exercise—The Best Ice Cream Cone

2. Exploration Activity—A Rational Domain

3. Conceptual Exercise—Rational Fuel Efficiency

4. Group Activity—Counting Fish

5. Graphing Calculator Activity—Travel Time

1. Extension Exercise—The Best Ice Cream Cone

Suppose an ice cream cone company wants to know the best size to make their sugar cones if the volume must stay fixed at $60\,\text{cm}^3$. The most economical approach would be to use the least amount of ingredients. Therefore, they need to minimize the surface area of a cone.

a. The formula for the volume of a cone is given below. Substitute $60\,\text{cm}^3$ for the constant volume V to obtain an equation in terms of the radius r and height h. Then isolate h on one side of the equation so that the height h is written in terms of the radius r.

$$V = \frac{1}{3}\pi r^2 h$$

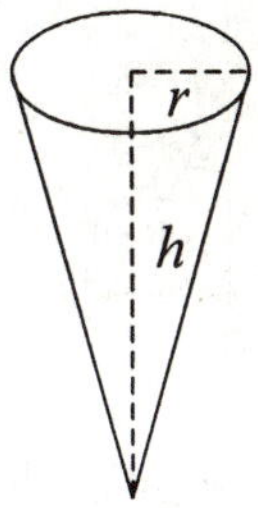

b. The formula for the surface area of a cone is given below. To develop an equation for surface area in terms of radius r, substitute the expression obtained in part *a* for height h.

$$A = \pi r \sqrt{r^2 + h^2}$$

c. Use the formulas developed in parts *a* and *b* for height and surface area to complete the table below. Round results to the tenths place. Observe the change in height and surface area as the radius increases.

Radius (cm)	Height (cm)	Surface Area $\left(\text{cm}^2\right)$
0.5		
1		
1.5		
2		
2.5		
3		
3.5		
4		
4.5		
5		

d. Graph A as a function of r to see how surface area depends on radius.

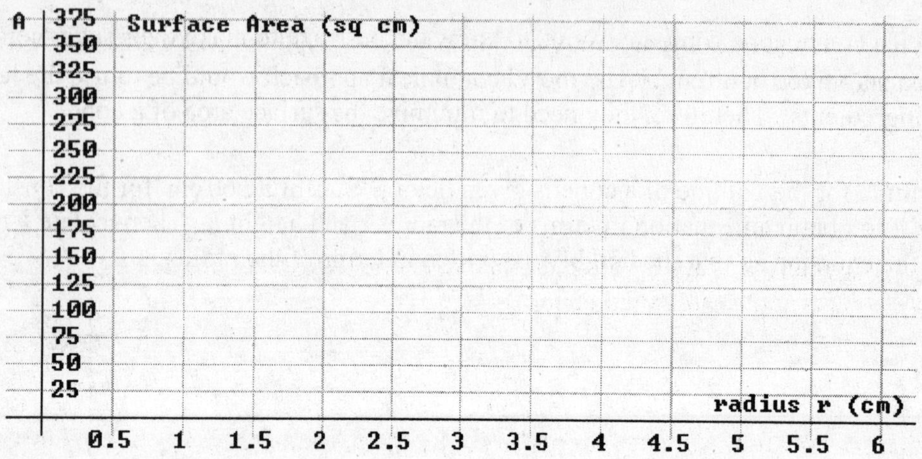

e. Using the table from part *c* and the graph from part *d*, estimate the minimum surface area to the nearest square centimeter. Then write the corresponding radius and height to the nearest tenth of a centimeter.

f. Now create a second table that shows the radius from 3 cm to 4 cm in steps of 0.1 centimeter and the corresponding height and surface area. Round the height and surface area to the hundredths place.

Radius (cm)	**Height (cm)**	**Surface Area** (cm^2)
3		
3.1		
3.2		
3.3		
3.4		
3.5		
3.6		
3.7		
3.8		
3.9		
4.0		

g. Use the table created in part *f* to make a more accurate estimate of the minimum surface area and its corresponding radius and height.

h. How practical will the cone size be for eating ice cream?

2. Exploration Activity—A Rational Domain

The function below represents yearly savings in the cost of car fuel by increasing gas mileage from x miles per gallon to $x + 5$ miles per gallon.

$$f(x) = \frac{112{,}500}{x(x+5)}$$

a. What is the domain of f if we ignore the problem situation and assume that x does not represent any particular unknown quantity?

b. What is the practical domain of f in terms of the problem situation? Let x be the current gas mileage in miles per gallon.

c. Complete the table by entering values for $y = f(x)$ in the second column.

Current Gas Mileage x **(miles/gallon)**	**Savings in Fuel Cost** y **(dollars)**
5	
10	
15	
20	
25	
30	
35	
40	
45	
50	
55	
60	

d. Observe the table for patterns, then complete the following sentence.

As current gas mileage increases, ______________________________

____

e. Graph the function *f* over the practical domain (from part *b*).

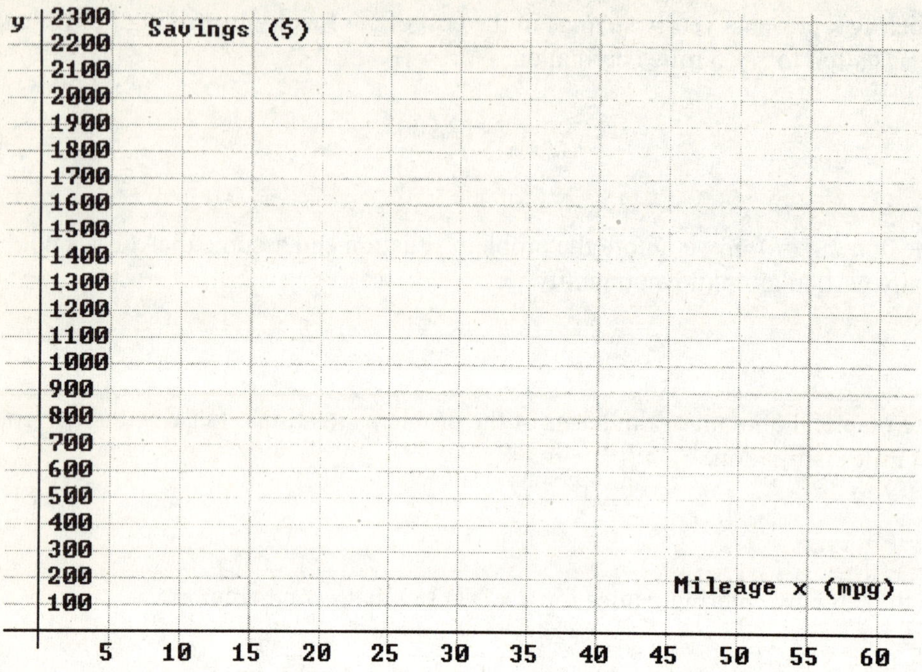

f. What happens to the function *f* as $x \to \infty$ (*x* approaches infinity, or *x* gets larger and larger)?

g. What happens to the function *f* as $x \to 0^+$ (*x* approaches 0 from the right, or *x* gets smaller and smaller toward zero from the positive direction)?

h. What should a consumer know about the relationship between savings and gas mileage when looking to buy a car?

3. Conceptual Exercise—Rational Fuel Efficiency

So you want to buy a car and need to find out how much money you can save by improving gas mileage. Suppose that you drive 18,000 miles per year, pay an average of \$1.25 per gallon, and currently own a car that gets only 10 miles per gallon. Your goal is to develop a rational function that models savings in fuel cost from a 5-mile per gallon increase in fuel efficiency.

a. How much do you pay for fuel per year? *Hint:* Use the formula given below.

$$\text{Cost} = \frac{\text{miles/year}}{\text{miles/gal}} \times (\text{price/gal})$$

b. If you buy a new car that gets 15 miles per gallon, how much will you pay for fuel per year? Show all work, and express each number with its proper unit.

c. How much money is saved by improving gas mileage from 10 miles/gallon to 15 miles/gallon? Show all work.

d. To generalize the work done in parts *a* – *c*, let x be your current gas mileage in miles per gallon (instead of 10). Write a rational expression that represents annual fuel cost in terms of x. *Hint:* Refer to part *a* but use x in place of 10.

e. If x represents current gas mileage, then how do you express a 5-mile/gallon increase in gas mileage in terms of x?

f. Now write a new rational expression to represent annual fuel cost after increasing gas mileage by 5 miles/gallon. *Hint:* Refer to part *d* but use $x + 5$ in place of x.

g. Let y be the annual savings in fuel cost by improving gas mileage by 5 miles/gallon. Find an equation for y in terms of x by taking the difference of the rational expressions in parts *d* and *f*. *Hint:* Use the process outlined below.

Annual savings = (Expression in part *d*) – (Expression in part *f*)

h. Simplify the equation for y in part g by combining the two rational expressions into one. *Hint:* Find a common denominator.

i. If $y = f(x)$, express your equation for y (from part h) using function notation.

j. Find the value of $f(10)$ and explain its meaning in terms of the problem situation.

4. Group Activity—Counting Fish

It has been five years since ChemDump Inc. was forced to clean up their act and stop polluting a stream that runs into Crystal Lake. At that time, the Tabasco trout and Loudmouth bass were at an all-time low in numbers, but now the local fishermen think the fish are on the rise again. Five years ago, the EPA took a count of the two populations of fish using a capture-tag-recapture procedure. You want to try this strategy now to see if there really is a significant increase in the fish population.

The EPA performed the procedure by catching fish in a large net, tagging each trout and bass, then releasing them back into the lake to mingle with the whole population. A week later they made a second catch in which they examined the total number of trout and bass and counted the number of recaptured (tagged) fish. The data is listed in the table below.

Fish	Tagged in First Catch	Total Examined in Second Catch	Recaptured (Tagged) in Second Catch
Trout	196	142	18
Bass	255	271	23

a. Let t be the total number of trout in Crystal Lake and b the total number of bass. For each fish listed, what is the ratio (fraction) of "tagged in first catch" to total in lake?

Trout: Bass:

b. For each fish listed, what is the ratio (fraction) of "recaptured" to "total examined in second catch?"

Trout: Bass:

c. Use information from parts *a* and *b* to set up and solve a proportion that estimates the total number of trout t in Crystal Lake. Round to the nearest fish.

d. Use information from parts *a* and *b* to set up and solve a proportion that estimates the total number of bass b in Crystal Lake. Round to the nearest fish.

e. What do your results mean in terms of the problem situation?

f. Explain why the following strategies would not work in our problem situation.

- The second catch is made a couple of hours after the first catch, in the exact same location.

- The second catch is made one year after the first catch.

g. If the capture-tag-recapture procedure is to work, we have to assume that what happens to the sample of living creatures that are caught, tagged, and released between the first and second catch?

5. Graphing Calculator Activity—Travel Time

The driving distance between Boston to New York City is 206 miles. Before driving that distance, you want to know how long it will take to reach your destination traveling at different speeds.

a. Complete the third column to see how travel time depends on the car's average speed or rate in miles/hour. Use the given formula and round your answer to the nearest tenth of an hour. The first two times have already been filled in.

$$\text{time} = \frac{\text{distance}}{\text{rate}} \quad \text{or} \quad t = \frac{d}{r}$$

Distance d (miles)	Average Speed or Rate r (miles/hour)	Travel Time t (hours)	Change in Travel Time Δt (hours)
206	10	20.6	
206	20	10.3	$10.3 - 20.6 = -10.3$
206	30		
206	40		
206	50		
206	60		
206	70		
206	80		
206	90		

b. Explain how average speed (rate) affects travel time by completing the following sentence.

As the average speed increases, ______________________________

____

____

c. Complete the fourth column, then compare the change in travel time from 10 to 20 mph, from 20 to 30 mph, and from 30 to 40 mph. What general statement can you make about increasing rate when traveling at slower speeds?

____

____

____

d. Now compare the change in travel time from the following changes in speed: 60 to 70 mph, 70 to 80 mph, and 80 to 90 mph. What general statement can you make about increasing the rate when traveling at faster speeds?

____

____

____

e. Write a function $t(r)$ that expresses how time t depends on rate r. *Hint:* Assume that distance is a constant 206 miles.

$$t(r) =$$

f. Use your function from part *e* to find the time it will take to drive from Boston to New York if your average rate is 65 mph. Use function notation and show the necessary calculations. Write your answer in hours and minutes, rounding minutes to the nearest whole minute.

g. If the trip takes 3 hours and 45 minutes, then what is your average speed? *Hint:* Change 45 minutes to the correct fraction of an hour so that time is expressed in hours. Then set up and solve the equation. Round your answer to the nearest mile per hour.

h. If it takes 7½ hours (no stopping) to travel from Boston to Washington D.C. at an average rate of 40 mph, find the travel time when the average rate is 50 mph. *Hint:* First use the given information and the formula $d = rt$ to find the distance d.

i. If we travel from Boston to Washington D.C., write a function that models how the time t of this trip depends on the average rate (speed) r.

Chapter 7—Rational Exponents, Radicals, and Complex Numbers

1. Extension Exercise—The Pendulum

2. Exploration Activity—Driving Conditions

3. Conceptual Exercise—Tractor Pull

4. Group Activity—Maximum Volume of a Cone

5. Graphing Calculator Activity—The Square Root of a Square

1. Extension Exercise—The Pendulum

Make a simple pendulum by tying a string tightly to a weight. The formula relating a pendulum's period T (in seconds) to its length L (in centimeters) is $T = 2\pi\sqrt{\frac{L}{980}}$.

The period of a pendulum is defined as the time it takes the pendulum to complete one full back-and-forth swing. In this activity, you will be measuring your pendulum's period with a stopwatch. Because the periods will be only a few seconds long, it will be more accurate for you to measure the time for a total of five complete swings and then find the average time of one complete swing.

Materials needed: string (at least 1 meter long), weight, meter stick, stopwatch, calculator

a. For each of the pendulum (string) lengths given in the table below, measure the time required for 5 complete swings and record it in the appropriate column. Next, divide this value by 5 to find the measured period of the pendulum for the given length and record it in the third column of the table. Use the given formula to calculate the theoretical period T for the same pendulum length and record it in the fourth column (rounded to two decimal places). In the last column, find and record the absolute value difference between the measured period and the theoretical period.

Length L (cm)	**Time for 5 Swings (seconds)**	**Measured Period T_m (seconds)**	**Theoretical Period T (seconds)**	**Difference $\lvert T - T_m \rvert$**
30				
55				
70				

b. For each period T in the table below, use the given formula $T = 2\pi\sqrt{\frac{L}{980}}$ to calculate the theoretical pendulum length L required. Record L in the second column (rounded to one decimal place).

Next, use this length L for the string pendulum. Measure and record the time for 5 complete swings. Divide this value by 5 to find the measured period T_m, and record it. Then find and record the absolute value difference between the theoretical period and the measured period.

Period T (seconds)	**Theoretical Length L** (centimeters)	**Time for 5 Swings** (seconds)	**Measured Period T_m** (seconds)	**Difference** $\lvert T - T_m \rvert$
1				
1.25				
2				

c. Use the general trends you find in the tables from parts *a* and *b* to describe the relationship between a pendulum's period and its length.

____

____

____

d. Write a paragraph explaining the differences you found between the values of the theoretical period and the measured period. Explain what factors contributed to these differences.

____

____

____

____

____

____

2. Exploration Activity—Driving Conditions

When investigating car accidents, police departments find it useful to approximate the speed of a car when given the distance that the car skidded before it came to a stop.

The function, $S(x) = \sqrt{30kx}$, is used to estimate the speed S (miles per hour) that a car was traveling initially if the car stopped after leaving skid marks of length x. The value of k depends on the type of pavement and the road conditions. The table below gives values of k for four types of pavement and road conditions.

Value of *k*	**Pavement/Road Conditions**
$k = 0.8$	Dry concrete
$k = 0.35$	Wet concrete
$k = 1.0$	Dry tar
$k = 0.44$	Wet tar

a. Substitute the appropriate value for k to write the speed S as a function of x for each of the road types and conditions.

Pavement/Road Conditions	***S(x)***
Dry concrete	$S(x) =$
Wet concrete	$S(x) =$
Dry tar	$S(x) =$
Wet tar	$S(x) =$

b. Use the formulas from part *a* to determine the approximate speed that a car was traveling for several lengths of skid marks on different surfaces. Fill in the following table with the results, rounding to the nearest mile per hour.

Length (ft) of Skid Mark, x	Speed (mph) on Dry Concrete, $S(x)$	Speed (mph) on Wet Concrete, $S(x)$	Speed (mph) on Dry Tar, $S(x)$	Speed (mph) on Wet Tar, $S(x)$
150				
200				
250				
300				

c. Use a graphing calculator to plot the four curves for $S(x)$ for the four different road conditions. Show the graphs on the grid below. Use four different colors, or types of lines to distinguish the four different curves. Clearly label which curve represents which road type and condition.

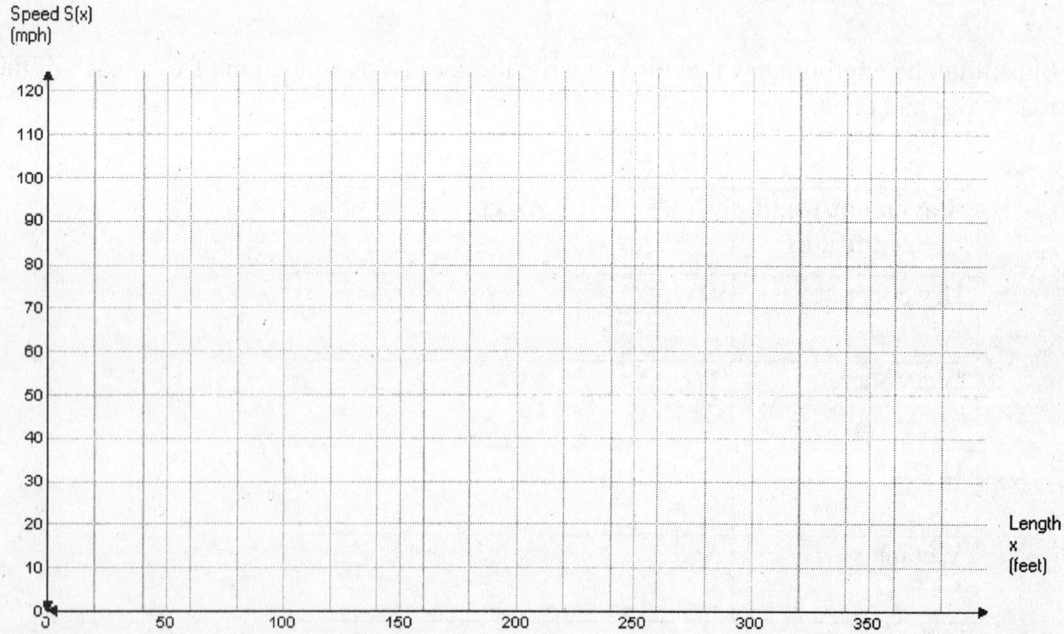

d. From the graph in part *c*, determine the approximate stopping distance of a car traveling at the following speeds for each condition.

Speed (mph)	**Stopping Distance Dry Concrete** (ft)	**Stopping Distance Wet Concrete** (ft)	**Stopping Distance Dry Tar** (ft)	**Stopping Distance Wet Tar** (ft)
30				
65				
100				

e. The safest road condition occurs when a car going a particular speed can stop in the shortest distance. From the data and graphs, which road condition appears to be the safest and which appears to be the most dangerous? Explain how you made your choices.

____

____

____

____

____

f. What other factors might determine the length of skid marks a car leaves when trying to stop suddenly?

____

____

____

____

3. Conceptual Exercise—Tractor Pull

Two tractors are pulling a tree stump from a field. If two forces, *A* and *B*, pull at right angles (90°) to each other, the size of the resulting force *R* is given by the formula $R = \sqrt{A^2 + B^2}$.

The forces of the two tractors and the resulting force can be modeled mathematically as the two legs and hypotenuse of a right triangle. The resulting force is computed using the Pythagorean theorem. *Hint:* The rectangle below is divided into two right triangles, each with legs *A* and *B*.

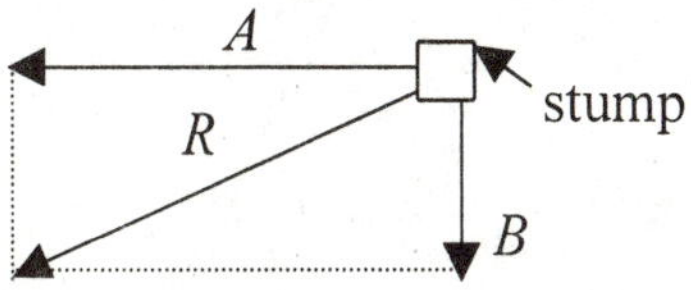

$R = \sqrt{A^2 + B^2}$

Suppose that tractor *B* pulls with twice as much force as tractor *A*. Fill in the chart below to see what happens if tractor *A* pulls with the forces shown in column 1 of this chart.

In the last column, write ordered pairs where the first coordinate is the force exerted by tractor *A*, and the second coordinate is the resultant force *R*.

a.

Force Exerted by Tractor *A* (pounds)	**Force Exerted by Tractor *B*** (pounds)	**Resultant Force** *R* (nearest tenth of a pound) $R = \sqrt{A^2 + B^2}$	**Ordered Pairs *(A , R)***
100			
200			
300			
400			
500			
600			

b. Express the resultant force *R* as a function of *A*. First express *B* in terms of *A,* then substitute the expression for *B* into the expression for *R*. Write the *R* formula in expanded form and simplify the radical. *Hint:* Remember that *A* is always a positive number in this activity.

R(A) = ________________________ = ________________________

expanded   simplified

c. Use a graphing calculator to graph the simplified formula from part *b*. Pick an appropriate window to represent the data. Use your tracing function to confirm that the graph goes through the points plotted. In the space below, write the coordinates of the six points you get from tracing the function. Due to rounding, there may be slight differences in coordinates.

(,) (,) (,)

(,) (,) (,)

d. Your graph should be a straight line. Explain how you could have predicted (from part *b*) that your graph would be a straight line.

____

____

____

e. Set up an equation to determine the force tractor A must exert for the resultant force R to be 250 pounds. Solve the equation and express the solution exactly (using a radical). Then round that solution to the nearest pound. Write the coordinates of the ordered pair on your graph (from part *c*) that is associated with the solution.

Equation: ____________________

Solution: ____________________ $\approx$ ____________________
exact approximate

Ordered pair associated with the solution: (,)

4. Group Activity—Maximum Volume of a Cone

In this activity, you will determine what size wedge (shaped like a piece of pie) to cut from a circle to form a cone of the largest possible volume.
Materials needed: paper, string, scissors, tape, ruler

a. From paper, cut four circles, each with a 5-inch radius. Compute the circumference (rounded to the nearest tenth) of each circle. *Hint:* $c = 2\pi r$ is the formula for computing the circumference c from its radius r. If you do not have a compass, hold one end of a string at the center, tie a pencil point to position 5 inches from the center, and rotate the pencil point around to draw the circle.

$c \approx$ ____________ inches

b. Using string to measure, mark the following four lengths x around each rim of the four circles: x = 4", 6", 8", 10". For each of the circles, draw a wedge from the center of the circle with rim length x. Cut out the resulting wedge. Form what is left of each circle into a cone. Since the base of the cone is a circle, the radius and circumference are related by the equation, $C = 2\pi R$. Solve this equation for R in terms of C.

$R =$ __________________________

c. Compute the circumference C and radius R of the base of the four cones. Round results for C and for R to the nearest tenth. Capital letters are used to distinguish the circumference and radius of the base from the original circles.

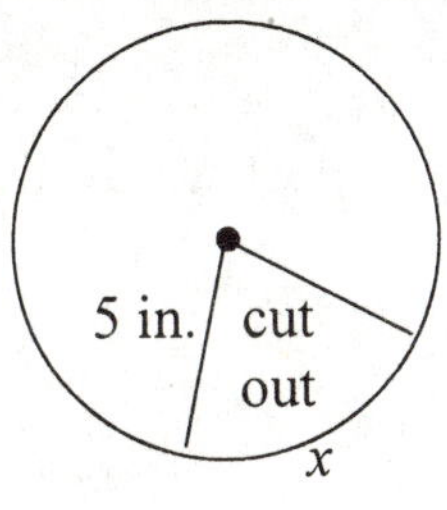

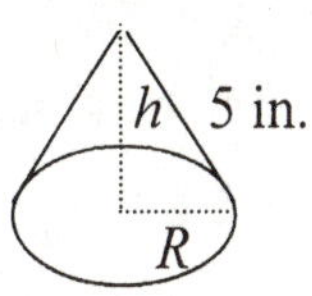

Length curved edge, x, (on the wedge)	**Circumference of base of Cone** $C = c - x$ (inches)	**Radius of Base of Cone, R:** $R = \frac{C}{2\pi}$ (in.)
x = 4 inches		
x = 6 inches		
x = 8 inches		
x = 10 inches		

The vertical cross section through the vertex of each cone forms a right triangle with base R and height h. The Pythagorean theorem gives the equation, $h^2 + R^2 = 25$. Solving the equation for h gives: $h = \sqrt{25 - R^2}$.

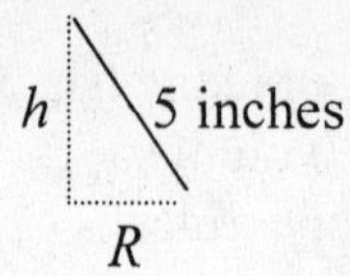

The volume of a cone is given by the formula, $V = \frac{1}{3}\pi R^2 h$. Substituting the above expression for h into the V formula, we get, $V(R) = \frac{1}{3}\pi R^2\sqrt{25 - R^2}$

d. Using the values for R from part c, compute the volumes (rounded to the nearest tenth) of your four cones. Record your results in the table below.

x (length on the rim of wedge cut out)	**Radius, R, of Base of Cone** (inches)	**Volume of Cone** (cubic inches)
4 inches		
6 inches		
8 inches		
10 inches		

e. Of the four cones you made, which one had the largest volume? In other words, the cone with the largest volume had what value for x ?

$x =$ ____________ Volume = ____________

f. If the cone of largest volume were formed from a circle with a radius of five inches, would the wedge cut out have one of the values for x in part e or would the x-value be between values in part e ? Discuss this question with your group and record your conclusions below.

____

____

____

____

5. Graphing Calculator Activity—The Square Root of a Square

There are two square roots of any positive number: one is negative, and the other is positive. When using the square root symbol, $\sqrt{\ }$, we are referring to the *positive* or *principal* square root.

a. For each of the numbers in Column 1 of the table, compute the square of the number and record your answer in Column 2. Then in Column 3, record the principal square root of the number.

x	**x^2, the square of x**	**principal square root of x^2**
5		
11		
-7		
15		
-15		
-14.3		
221		
-296		
$-\pi$		

b. Write a single mathematical function that takes you through the steps you went through above to get the answer in Column 2, then in Column 3.

$f(x) =$ ____________________________________

c. Compare the numbers in Column 3 with the original numbers in Column 1. Write a simpler function (than in part *b*) to give you the same numerical result as in part *b*.

$g(x) =$ ____________________________________

d. Plot both functions, *f(x)* and *g(x)*, on your graphing calculator to check that the two functions are identical. If they are not identical, double-check your work. Sketch *f(x)* and *g(x)* below.

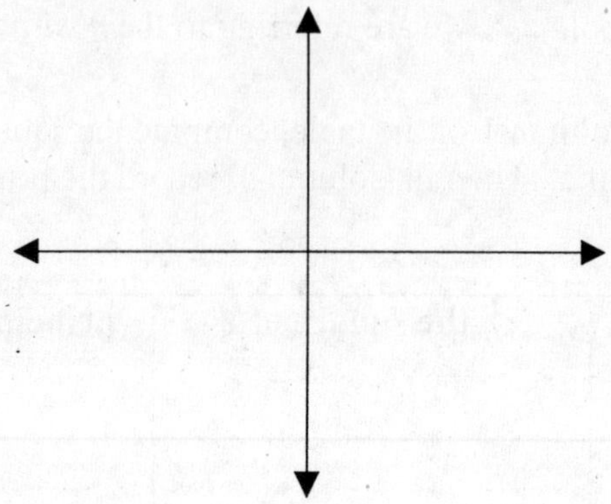

e. Explain or show why the following statement is *false*. $\sqrt{x^2} = x$

f. In light of the observations above, how would you simplify $\sqrt{x^2}$?

$\sqrt{x^2} =$ ____________________

g. Based on your work in parts *a* through *f*, determine which statement is true and which is false. Explain how you decided.

$\sqrt{(3x+5)^2} = 3x + 5$ True False

$\sqrt{(3x+5)^2} = |\ 3x + 5\ |$ True False

Chapter 8—Quadratic Equations and Functions

1. Extension Exercise—Ball Drop

2. Exploration Activity—Freefall

3. Conceptual Exercise—The Golden Ratio

4. Group Activity—A Hanging Cable

5. Graphing Calculator Activity—Quadratic Graphs

1. Extension Exercise—Ball Drop

April is standing on the edge of a cliff that overhangs a river. She wants to know how high the cliff is above the water's surface, so she drops a tennis ball and clocks how long it takes for the ball to hit the water. Four seconds after dropping the ball, she sees it hit the water. April knows that the distance *d* (in feet) that the ball is above ground (above the river) *t* seconds after being dropped can be calculated by the formula,

$$d = s(t) = -16\,t^2 + s_0$$

where s_0 is the height above the river from which the ball is dropped.

April's first task is to determine the value of s_0, the height of the cliff above the river.

a. Since $s(t)$ is the height of the ball above the river after t seconds, use the given information and the meaning of $s(t)$ to determine the values of $s(t)$ and t when the ball first hits the water/ Explain how you determined these values.

$s(t) =$ ____________ feet $t =$ ____________ seconds

____

____

____

____

b. Substitute the values from part *a* for $s(t)$ and t into the equation, $s(t) = -16t^2 + s_0$, and solve the resulting equation for s_0 to find out how high the cliff is above the river.

Equation after substitutions: ________________________________

Solution: $s_0 =$ ____________ feet above the river.

c. Substitute the value for s_0 from part *b* into the formula, $s(t) = -16\,t^2 + s_0$. Write the resulting function below.

$s(t) =$ ________________________________

d. Graph the parabola defined by $s(t)$ for values of t from -5 to 5.

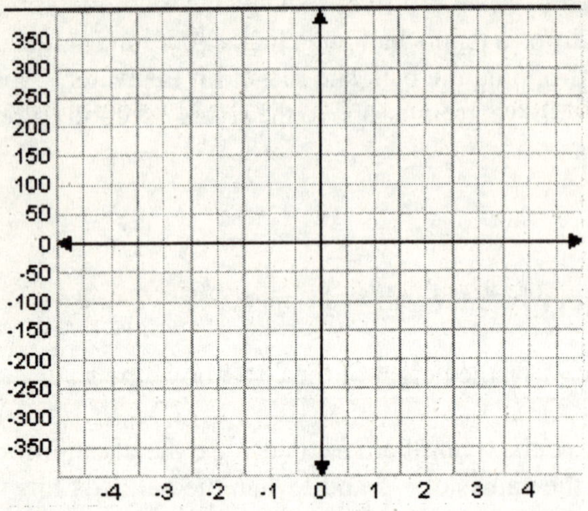

e. For each of the following situations, set up and solve an appropriate equation to find out how long it takes the ball to reach each of the given heights (above the water) and how far the ball has fallen when it reaches that height. Write an ordered pair for each situation and plot it on the graph above. Disregard the negative solutions. Round results to the nearest tenth.

40 feet above the water: distance fallen: ______________

equation: ______________________ ordered pair: ______________

time of fall: ______________________

The ball is half-way down the cliff: distance fallen: ______________

equation: ______________________ ordered pair: ______________

time of fall: ______________________

f. Although the graph in part *d* is shown for values of t from -5 to 5 and given the physical restraints of the function $s(t)$ in the context of the falling ball, what is the domain and range of the function that describes the falling ball? Explain your answer.

____

____

____

2. Exploration Activity—Freefall

The quadratic model that describes the position of a free falling object as a function of time (neglecting air resistance) is $\frac{1}{2}gt^2 + v_0t + s_0$, where t is time in seconds, s_0 is the object's initial height or position above ground in feet or meters, v_0 is the object's initial velocity in feet per second or meters per second, and g is the acceleration due to gravity in feet per second per second or meters per second per second.

If an object is thrown or launched *upward*, the initial velocity v_0 is *positive*. If the object is thrown or launched *downward*, v_0 is *negative*. Since gravity is a downward pull, the value for g is negative. The table below shows values for g on earth as well as on the moon.

Place	***g*** (feet per second per second)	***g*** (meters per second per second)
Earth	-32	-9.8
Moon	-5.28	-1.617

a. A rock is thrown downwards from the top of a 110-foot cliff at a velocity of 15 feet per second. Write functions that describe the position of the rock after t seconds if the cliff is on Earth and if the cliff is on the moon. Graph each function on the grids below. Label the scales you used on the grids.

On the moon: $f(t)$ = ____________________ On Earth: $f(t)$ = ____________________

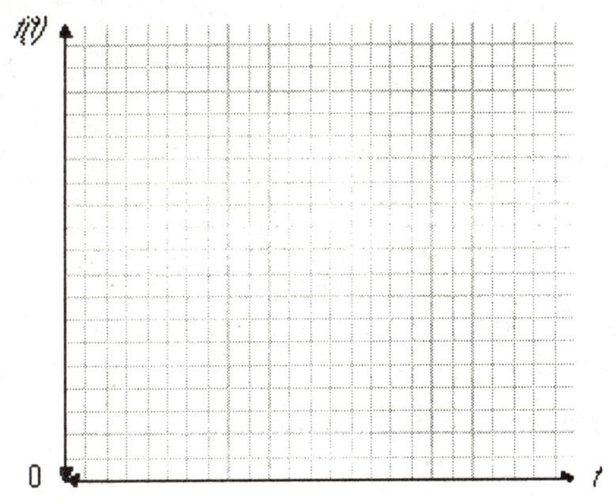

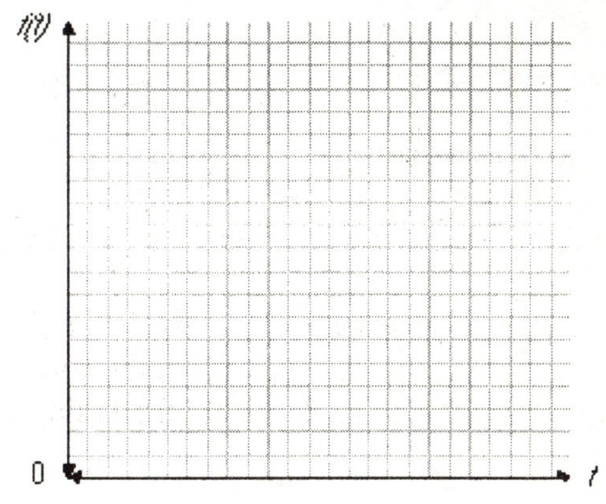

b. Does the rock hit the ground sooner on the moon or on Earth? Explain your answer.

____

____

c. A golf ball is thrown upward at a velocity of 100 feet per second from a height of 5 feet on the moon. Another golf ball is thrown upward at a velocity of 30.48 meters per second from a height of 1.524 meters on the moon. Write functions that describe the position of each golf ball after t seconds. Graph each function on the grids below.

100 ft/s from 5 ft:
$f(t) =$ ______________________

30.48 m/s from 1.5424 m:
$f(t) =$ ______________________

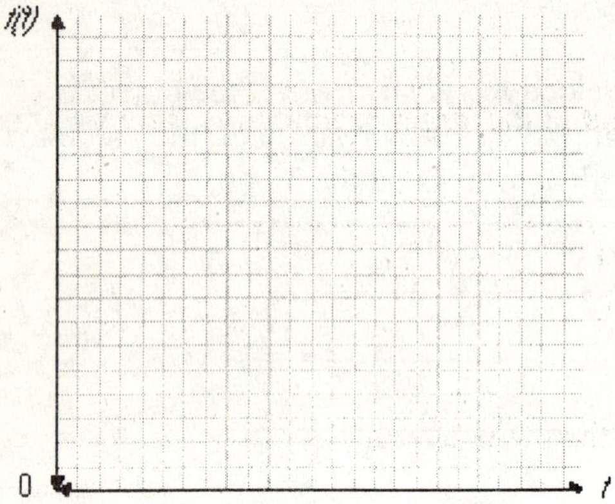

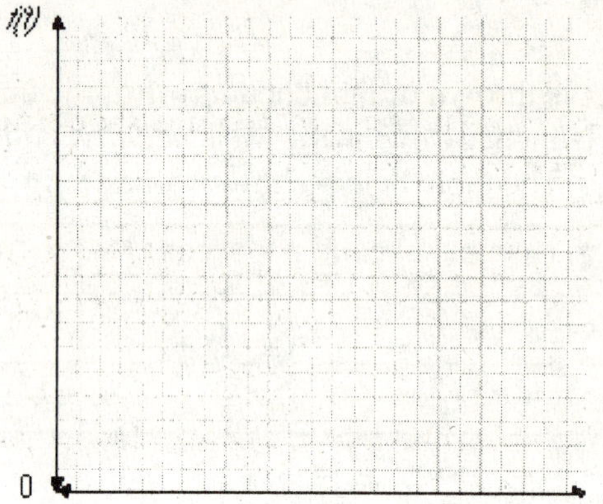

d. Use the graphs in part *c* to answer the following questions. What is the maximum height the ball reaches in each situation? How many seconds does it take to reach the high point? How long does it take for the ball to reach the ground? Round each result to the nearest hundredths place.

100 ft/s from 5 ft:

vertex: ______________________

maximum height: ______________________

time to reach maximum height: __________

time to reach ground: ______________________

30.48 m/s from 1.524 m:

vertex: ______________________

maximum height: ______________________

time to reach maximum height: __________

time to reach ground: ______________________

e. Explain why it takes the same amount of time to reach the maximum height in each situation and the same amount of time to reach the ground in each situation. *Hint:* 1 foot = 0.3048 meter.

____

____

____

3. Conceptual Exercise—The Golden Ratio

Some numbers and ratios appear many places in nature, art, and in mathematics. One such ratio is called "the golden ratio." The golden ratio can be defined mathematically, but it also appears in nature as a number that defines the type of spiral in a snail's shell or the pattern of sunflower seeds growing on a sunflower. It is common in classical art and architecture as well. In this activity, you will find the mathematical value of the golden ratio.

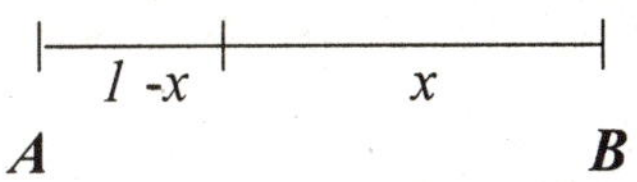

Length from A to $B = 1$

a. Write the ratio (in terms of x) of the shorter segment to the larger segment (as shown on the above diagram).

________________ (shorter segment) / (longer segment)

b. Write the ratio (in terms of x) of the longer part to the entire length (1 unit).

________________ / 1 (longer segment) / (one)

c. Set the two ratios from parts *a* and *b* equal to each other to form an equation in terms of the variable x.

d. To clear the fractions, multiply by x on both sides of the equation in part *c*. Solve the resulting quadratic equation. Use the quadratic formula and leave your answer in radical form. Show your work and record both solutions below.

$x =$ ____________________ or $x =$ ____________________

e. One of the solutions in part *d* is positive and one is negative. Use a calculator to find a decimal approximation (rounded to 6 decimal places) of the *positive* solution. The positive solution is the "golden ratio" and is commonly represented by the Greek letter phi, ϕ.

$\phi \approx$ ____________________

f. Use your reciprocal key on your calculator ($1/x$ or x^{-1}) to calculate the reciprocal of the golden ratio. Then use your calculator to find a decimal approximation for the negative root in part *d* and calculate the golden ratio. Record your results below, correct to six decimal places. Explain how the two numbers are related.

The negative root: ____________ $\frac{1}{\phi} \approx$ ____________________

____

____

g. The golden ratio has been considered to be magic or divine in some cultures. For the special value ϕ, it is true that $\frac{1}{\phi} + 1 = \phi + 2$, $1 + \phi = \frac{1}{\phi}$,

and $2\phi = \phi^3 + 1$. To see this, approximate the expressions below to show that each expression on the left is equal to its corresponding expression on the right.

$\frac{1}{\phi} + 1 \approx$ ____________________ $\phi + 2 \approx$ ____________________

$1 + \phi \approx$ ____________________ $\frac{1}{\phi} \approx$ ____________________

$2\phi \approx$ ____________________ $\phi^3 + 1 \approx$ ____________________

h. Substitute the following numbers into the equations below to show that the above numerical properties for ϕ are not generally true for other numbers.

Property: $\frac{1}{x} + 1 = x + 2$ $x = 3$ ____________________

Property: $1 + x = \frac{1}{x}$ $x = 4$ ____________________

Property: $2x = x^3 + 1$ $x = -2$ ____________________

4. Group Activity—A Hanging Cable

Materials needed:

- heavy, flexible string
- thumb tacks and/or tape
- a sheet of graph paper

On a piece of graph paper, draw a y-axis and an x-axis. Tape or tack your sheet of graph paper to a wall, blackboard, or bulletin board. Tape or tack one end of the string near the upper left corner of the graph paper. Pick another spot on the string to tack to the upper right corner of the paper. Allow enough slack in the string so that the middle of the string hangs below the x-axis. Let the string hang naturally and trace the shape of the curve it makes.

a. Mark eight different points on the curve you traced and record their approximate coordinates. Let the length of 1 square be one unit.

(,) (,) (,) (,)

(,) (,) (,) (,)

b. Use the curve-fitting feature of your graphing calculator to find the quadratic function that best fits the curve.

$f(x) =$ ____________________

c. Plot the points from part *a* on your graphing calculator. In words, describe how well the curve "fits" the data points. In your explanation, give points on the curve of $f(x)$ that appear closest to the points on your drawing.

____

____

____

____

____

____

____

Repeat the activity on the previous page for a different length of string. This time, let the string hang so that it lies above the *x*-axis.

d. Mark eight different points on the curve you traced and record their approximate coordinates. Let the length of 1 square be one unit.

(,) (,) (,) (,)

(,) (,) (,) (,)

e. Use the curve-fitting feature of your graphing calculator to find the quadratic function that best fits the curve.

$f(x)$ = ______________________

f. Plot the points from part *d*. Do the points appear to be on the curve? Does the curve appear to be a good fit? Explain.

____

____

g. What is the vertex of the parabola from part *e*? How does it compare with the lowest point of the hanging string?

Vertex of $f(x)$ ______________ Low point on string: ______________

In general, when a cable is hung from two supports, the shape of the hanging cable is approximately a parabola. There may be some variance because of the lack of uniformity in the cable. A parabola as a structure is very strong. For this reason, hanging cables have been used in buildings and bridges for support.

h. In your group, name examples where a bridge is supported from above by cables, or buildings where the roof seems to be hanging from cables or some other parabolic shape. Make a list of your examples on the lines below.

____

____

____

____

5. Graphing Calculator Activity—Quadratic Graphs

Keep in mind the following rules as you work through this activity:

- A quadratic function can have the form: $f(x) = ax^2 + bx + c$ where $a \neq 0$.
- If $a > 0$, the graph (called a parabola) opens upward. If $a < 0$, the parabola opens downward.
- The vertex of the parabola is the highest point (maximum) if the parabola opens downward, or is the lowest point (minimum) if the parabola opens upward. The coordinates of the vertex are: $\left(\frac{-b}{2a}, f\left(\frac{-b}{2a}\right)\right)$

a. For each quadratic function in the first column below, record the value of *a* in Column 2, and determine whether the related parabola opens upward or downward. Then calculate the coordinates of the vertex (correct to the nearest tenth), and determine the values of *x* that make the function zero. These *x*-values or solutions are also called *roots*.

$f(x) =$	**Value of *a***	**Opens up or down?**	**Vertex Coordinates**	**Roots, x_1 and x_2** (round real roots to the nearest tenth)
$5x^2 - 3x - 9$				$x_1 \approx$ $x_2 \approx$
$-15x^2 - 11$				$x_1 \approx$ $x_2 \approx$
$14x^2 + 12x + 5$				$x_1 \approx$ $x_2 \approx$
$-8x^2 + 3x + 7$				$x_1 \approx$ $x_2 \approx$

b. Which of the functions above have real number roots and which have complex roots?

Functions with real roots: ______________________________

Functions with complex roots: ______________________________

c. On your graphing calculator, graph each of the functions in the table. Pick a window where the shape of the parabola as well as the x-axis is clearly seen. Use your "trace" feature to find the vertex and x-intercepts (if there are any) of each parabola. Record your results below, correct to the nearest tenth. Show a rough sketch of each graph, labeling the vertex and the x-intercepts if there are any.

$f(x) = 5x^2 - 3x - 9$

vertex: (_____, _____)

intercepts (if any): (_____, _____) (_____, _____)

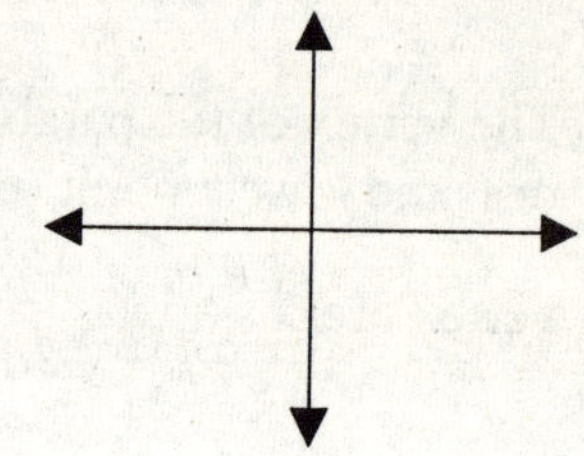

$f(x) = -15x^2 - 11$

vertex: (_____, _____)

intercepts (if any): (_____, _____) (_____, _____)

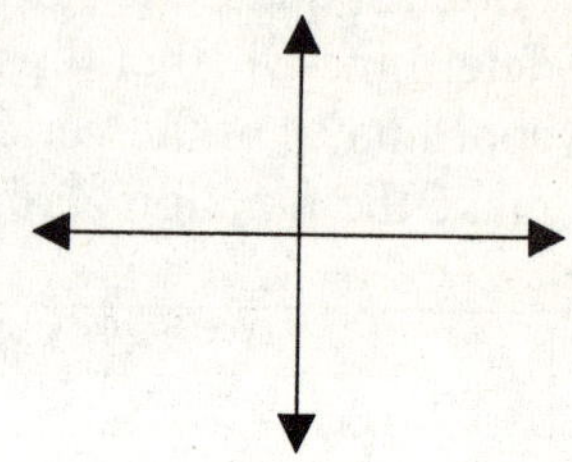

$f(x) = 14x^2 + 12x + 5$

vertex: (_____, _____)

intercepts (if any): (_____, _____) (_____, _____)

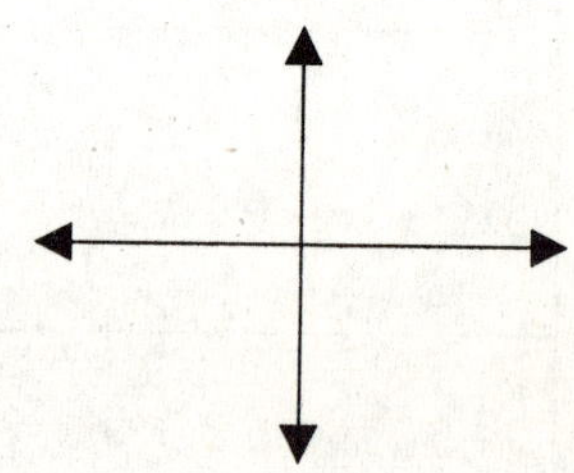

$f(x) = -8x^2 + 3x + 7$

vertex: (_____, _____)

intercepts (if any): (_____, _____) (_____, _____)

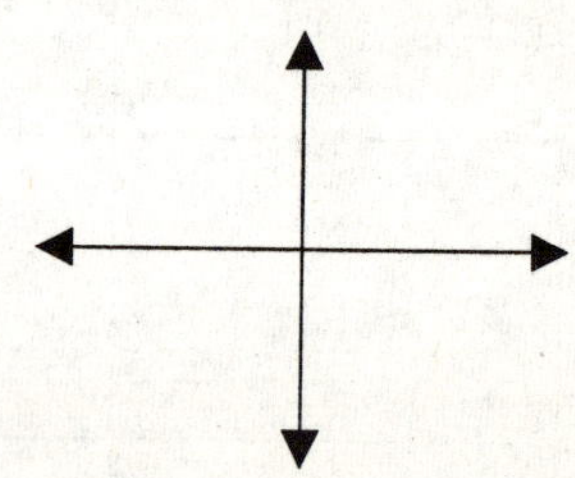

d. Explain the connection between the real number roots of a quadratic function and the x-intercepts of its graph. When a quadratic function has complex roots, what can be said about the x-intercepts of its graph?

____

____

____

____

Suppose $a > 0$ in $f(x) = ax^2 + bx + c$, and the vertex of the parabola is (-17, 4). In parts *e–h*, determine if the statement is true, false, or whether not enough information is available to make a decision. Show or explain how you decided.

e. The equation, $0 = ax^2 + bx + c$, has two real solutions.

True False Not enough info

f. The point (-17 , 4) is the highest point on the graph of $y = ax^2 + bx + c$.

True False Not enough info

g. $\frac{-b}{2a} = -17$ True False Not enough info

h. $f(5) = 22$ True False Not enough info

The graph of $y = ax^2 + bx + c$ has a vertex in the 2nd quadrant and opens down. In parts *i–k*, determine if the statement is true, false, or whether not enough information is available to make a decision. Show or explain how you decided.

i. The equation, $0 = ax^2 + bx + c$, has two real solutions. One is negative, and the other positive.

True False Not enough info

j. For the function $f(x) = ax^2 + bx + c$, $a > 0$.

True False Not enough info

k. The coordinates of the vertex of $f(x) = ax^2 + bx + c$ is (-2, 5).

True False Not enough info

Chapter 9—Conic Sections

1. Extension Exercise—The Space Race

2. Exploration Activity—Hyperbolic Skis

3. Conceptual Exercise—Parabolic Reflections

4. Group Activity—Fishing Area

5. Graphing Calculator Activity—Football Pass

1. Extension Exercise—The Space Race

Just as planets move in elliptical orbits with the sun at one focus, satellites can move in elliptical orbits with the center of the earth at one focus. In 1957 the Soviet Union sent into orbit the first artificial satellite, Sputnik I. During Sputnik's orbit, its altitude from the earth's surface ranged from a low of 132 miles to a high of 583 miles.

The figure below shows the orbit of Sputnik with the center of the earth as one of the foci so that $a + c$ is Sputnik's maximum distance from the earth's center, and $a - c$ is Sputnik's minimum distance from the earth's center. How can we use this information to find the equation that models Sputnik's orbit around the earth? *Hint:* The radius of the earth is 3950 miles.

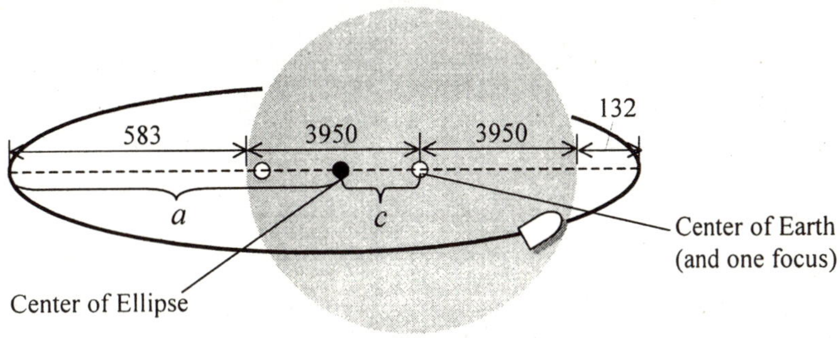

a. If the maximum distance from the earth's center is M and the minimum distance is m, then we have the following system of equations:

$$a + c = M$$
$$a - c = m$$

Substitute for M and m, then solve the system to find a and c.

b. Use $b^2 = a^2 - c^2$ to find b^2 and b. Round b to the tenths place.

c. We can see the nearly circular earth as if we are looking from deep space by graphing a circle with the origin as its center and a radius of 3950 miles.

$$x^2 + y^2 = (3950)^2$$

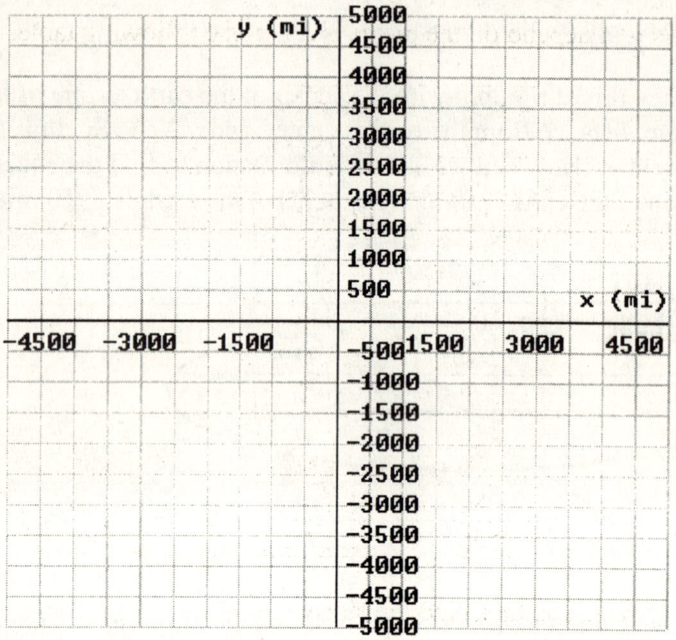

d. The orbit of Sputnik can be graphed as an ellipse with the center of the earth at one focus. Find an equation for Sputnik's orbit using the standard form,

$$\frac{(x-h)^2}{a^2} + \frac{(y-k)^2}{b^2} = 1$$

Assume that the x-axis is the orbit's major axis, the center of the earth is the origin and the right focus. Sputnik's orbit has center (h, k) located along the x-axis at a distance of c units to the left of the origin (0, 0).

e. Graph the equation for Sputnik's orbit around the earth on the xy-plane in part *c*.

f. Use the following formula to find the eccentricity e of Sputnik's orbit. Round to the ten-thousandths place.

$$e = \frac{c}{a}$$

2. Exploration Activity—Hyperbolic Skis

Suppose a ski manufacturer's research and development team has designed a new type of ski that has hyperbolic curves on both sides. The design allows the average skier to complete fancy turns with minimal effort. Based on a consumer survey, the marketing team has found that the annual demand for hyperbolic skis will depend on the prices given in the following table.

Price *p* ($)	**Demand *d* (skis/year)**
0	87,500
50	81,250
100	75,000
150	68,750
200	62,500
250	56,250
300	50,000
350	43,750
400	37,500
450	31,250
500	25,000

a. Complete the following sentence, then find the slope.

For each $50 increase in price the quantity demanded ______________________________

____

b. Explain the practical meaning of slope in terms of the problem situation.

____

____

c. Find the linear function in slope-intercept form that gives the demand d as a function of the price p.

d. Solve the equation found in part c for price p. Make sure that the equation is in slope-intercept form with p written in terms of d.

e. Total revenue R is calculated as price times demand (quantity sold),

$$R = p \cdot d$$

Find revenue R in terms of d by substituting the right side of the equation (found in part d) into the general revenue formula given above.

f. The research and development team estimates that the total cost of producing d units of hyperbolic skis is modeled by the following cost function C.

$$C = .002d^2 + 175d + 200{,}000$$

On the grid below, graph the revenue function found in part *e* and the cost function given above.

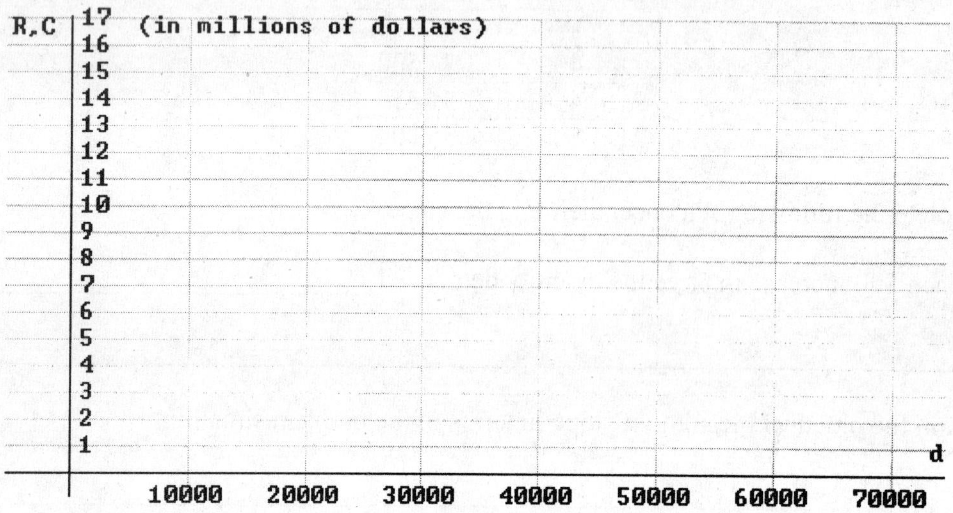

g. Over what interval will the demand for skis produce a profit greater than 0? In this interval, estimate the demand level where profit will be greatest. Explain your answer using complete sentences.

____

____

h. If profit is revenue minus cost, find the profit function for this situation and then graph the function in the coordinate plane for part *f.* Find the price that gives maximum profit by substituting this demand quantity into the price equation from part *d.*

3. Conceptual Exercise—Parabolic Reflections

A parabola can be defined as the set of all points that have an equal distance from a fixed point called the focus and a fixed line called the ***directrix***. Inside the head of a flashlight you will find a parabolic mirror which reflects the light shining from a bulb at the focus. The light coming from the focus strikes the parabola's edges and reflects out in lines that are parallel to the axis of symmetry. This is demonstrated in the figures below.

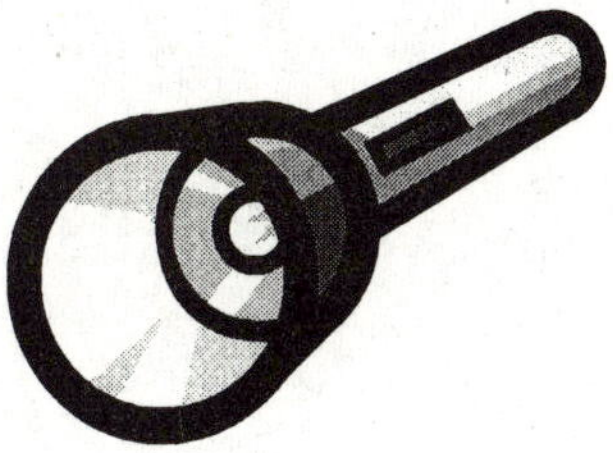

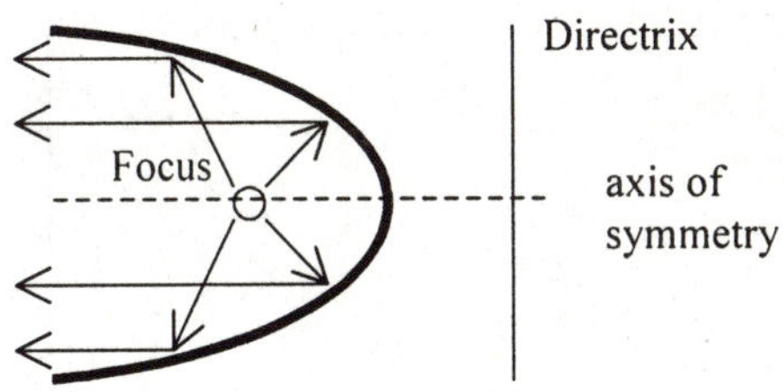

a. Suppose the mirror inside the flashlight is a parabolic solid with a diameter of 4 inches and a depth of 2 inches. The cross section is a parabola that can be modeled by the equation $x^2 = 2y$ on the interval [-2, 2]. Sketch this model below.

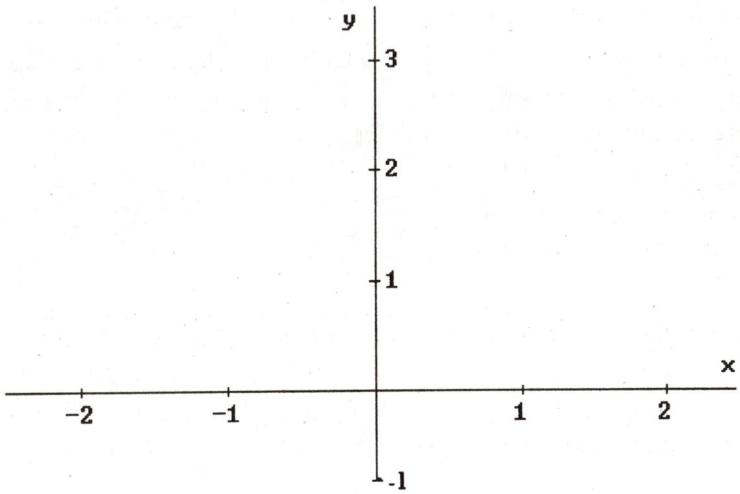

b. Parabolas in the standard form $x^2 = 4py$ have focus $(0, p)$ and directrix $y = -p$. Find the focus and directrix for the equation you graphed in part *a*, then sketch them on your graph.

c. Use the graph in part *a* to illustrate a light bulb at the focus emanating rays that bounce off the parabola's edge and reflect out of the parabola parallel to the axis of symmetry. What is the equation of the line that represents the axis of symmetry?

d. If you keep the diameter of the parabolic mirror at 4 inches but change the depth to 1 inch, then you can model a cross section with the equation $x^2 = 4y$. Draw a graph of this equation and label the point where the light bulb must be located in order for the flashlight to reflect light rays that are parallel to the axis of symmetry.

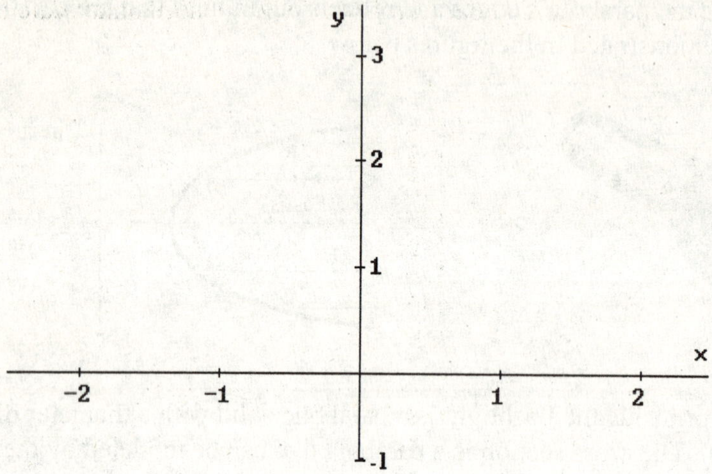

e. If you keep the diameter of the parabolic mirror at 4 inches but change the depth to 0.5 inches, then you can model a cross section with the equation $x^2 = 8y$. Draw a graph of this equation, and label the point where the light bulb must be located in order for the flashlight to reflect light rays that are parallel to the axis of symmetry.

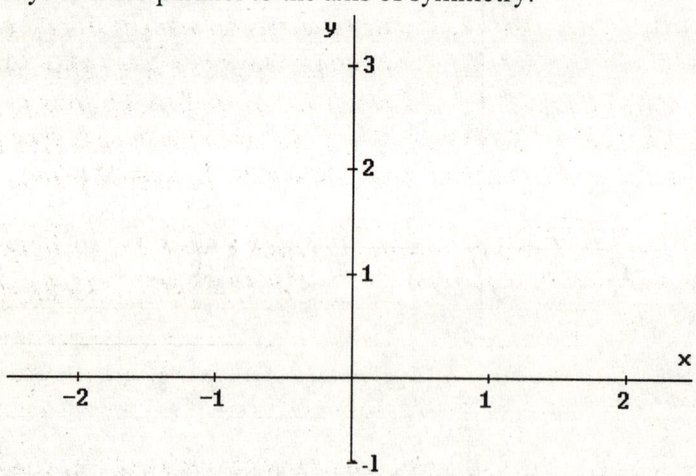

f. As you decrease the depth of the parabola, describe what happens to the location of the light bulb.

____

____

4. Group Activity—Fishing Area

Mike and his Dad like to row their boat out to the middle of Crystal Lake to try their luck at fishing. Mike has been after his Dad to buy a new rod and reel that will allow him to make longer casts in any direction and thus increase the area he has for fishing. For the graph below, suppose that the origin is the anchored boat and the vertical and horizontal scale measures distance in meters.

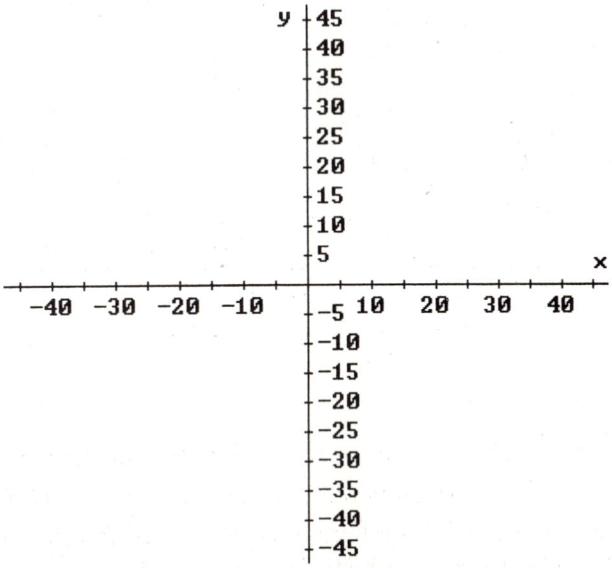

a. Dad takes Mike to Bud's Bait and Tackle where Bud has three types of rod/reel combinations. The table below lists the price and maximum casting radius (distance) in meters. Use the formula for the area of a circle $A = \pi \cdot r^2$ to find the amount of fishing area available to the nearest meter for each type. Then take the price of each type and divide by the fishing area to get the cost per square meter. Round to the nearest penny.

Rod/Reel Combo	**Price (dollars)**	**Casting Radius r (meters)**	**Fishing Area A (square meters)**	**Fishing Cost (dollars/m^2)**
Type 1	\$15.95	10		
Type 2	\$25.95	20		
Type 3	\$49.95	40		

b. In the *xy*-plane above, picture the fishing area available for each rod/reel combo by graphing the three circles that represent the maximum casting radius for each rod/reel type. Use (0, 0) for the center of each circle. If you double the radius do you get twice as much fishing area? Explain.

____

____

c. Write the equation of the circle that represents the maximum casting distance for each rod/reel combination type.

Type I: Type II: Type III:

d. Which type of rod/reel combination do you think is the best buy? Justify your answer using complete sentences.

____

____

____

____

e. Write the equation from part *c* that matches the "best buy" rod/reel type you decided on in part *d*. Solve for y in terms of x. Is y a function of x? Justify your answer using complete sentences.

Best Buy Equation:

____

____

____

f. What else should you consider before buying a rod/reel combination?

____

____

____

5. Graphing Calculator Activity—Football Pass

A quarterback warms up by throwing a football straight down the field in a trajectory path given by the quadratic equation,

$$y = -\frac{1}{20}x^2 + 1.1x + 6$$

where x is the horizontal distance in yards from where the ball is thrown, and y is the height of the ball in feet.

a. Use your graphing calculator to graph the flight path of the ball. Be sure that you obtain a window that is appropriate for the problem situation. Then sketch the graph below.

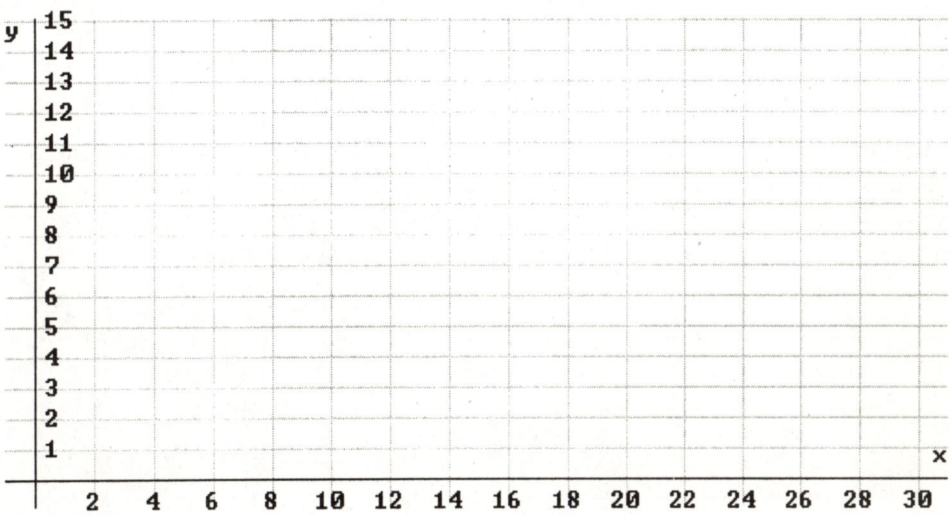

b. How high is the ball the instant it leaves the quarterback's hand?

c. Estimate the maximum height of the ball graphically using your calculator's trace or maximum feature. Then verify your answer algebraically using the vertex formula, $x = -\frac{b}{2a}$, and numerically using the calculator's table feature.

d. How far has the ball traveled if the receiver completely misses catching the ball and it hits the ground? Answer this question graphically using the calculator's trace or zero feature. Support your answer algebraically using the quadratic formula and then numerically using the calculator's table feature.

e. Using the same trajectory path, suppose that the quarterback throws the ball straight down the field to a 6-foot receiver standing 17 yards away in line with the path of the ball. Does the receiver have a reasonable chance to catch the ball? Explain your answer in complete sentences and describe the method you used to arrive at your answer.

___

___

___

___

f. Which coefficients would you change slightly so that the quarterback's trajectory will allow the receiver to catch the ball without jumping or reaching down? Assuming that the quarterback throws the ball overhand, what would your new equation look like?

Chapter 10—Exponential and Logarithmic Functions

1. Extension Exercise—Growth Limits

2. Exploration Activity—Growth Factor

3. Conceptual Exercise—Sweepstakes Winner

4. Group Activity—Simple Scales With Logarithms

5. Graphing Calculator Activity—The Value of Money

1. Extension Exercise—Growth Limits

The asexual reproduction of cells, known as *mitosis*, means that a cell can divide into two cells with each new cell having the same DNA as the original. This process allows identical genetic information contained in one cell to multiply again and again by a growth factor of two.

a. Assume that you have set up an experiment that places 100 cells under lab conditions that allow each one to divide into two cells every 20 minutes ($\frac{1}{3}$ of an hour). Fill in the table below to explore patterns in how this cell multiplies over time.

Time t (hours)	Cell Population
0	100
$\frac{1}{3}$	
$\frac{2}{3}$	
1	
$1\frac{1}{3}$	

b. Use the general exponential model given below to find a function that represents the growth of the cell population P with respect to time t. Note that P_0 is the initial population at $t = 0$, and $\frac{1}{k}$ is the time that it takes for the population to double.

$$P(t) = P_0 \cdot 2^{kt}$$

c. How many cells are in the population after 3 hours?

d. Suppose there are 409,600 cells when you leave for lunch. Upon returning you find that there are 3,276,800 cells. How long were you gone?

e. What would be the problem with trying to construct a graph that shows the population size over an 8-hour period?

____

____

____

f. Explain why an exponential growth model of any population whether it is cells, rabbits, or humans can only make accurate predictions within a restricted time interval.

____

____

____

____

g. Suppose an initial population of 1000 cells grows to 16,000 cells in two hours. What is the doubling time of this population? Use $P(t) = P_0 \cdot 2^{kt}$ to find k. Then recall that $\frac{1}{k}$ is the doubling time, or the time that it takes the population to double.

h. Suppose an initial population of 1000 cells grows to 17,000 cells in two hours. What is the doubling time of this population?

2. Exploration Activity—Growth Factor

The table below gives the total midyear population of Mexico from 1982 – 1988 rounded to the nearest thousand. Aside from the population P increasing each year, is there any pattern associated with how it is increasing?

Time t (year)	Population P	Change in Pop. ΔP	Growth Factor
0 (1982)	71,923,000		
1 (1983)	73,463,000		
2 (1984)	74,992,000		
3 (1985)	76,544,000		
4 (1886)	78,132,000		
5 (1987)	79,754,000		
6 (1988)	81,408,000		

(Source: U.S. Bureau of Census, International Data Base)

a. To get a clear idea of how the population is increasing, find the annual change in population ΔP. Place the result in the third column, and then complete the following sentence with either decreasing rate, constant rate, or increasing rate.

The population is increasing at a ________________________________

b. If a population is growing exponentially over a period of time, then there should be a constant growth factor each year—for example, the population increases by the same percentage every year. Divide each year's population by the previous year's population and put the result in the fourth column. Round your results to the thousandths place.

$$\text{Growth Factor} = \frac{\text{Current Year Population}}{\text{Previous Year Population}}$$

Explain what these values mean in terms of the problem situation.

____

c. What is the growth rate each year from 1982 – 1988? In other words, what is the constant percentage increase every year during this period of time?

d. In general, we can express the exponential growth of a population P as a function of time t using the model $P(t) = P_0 \cdot (b)^t$, where P_0 is the initial population at $t = 0$, and the base b is the constant growth factor (found in part b). Substitute the appropriate values for P_0 and b into the $P(t)$ model to obtain a formula for Mexico's population as a function of time.

$$P(t) =$$

e. Use the formula developed in part *d* to predict Mexico's population in the years 1990 and 1995. Use function notation and show the necessary computations to find each answer.

f. According to the U.S. Bureau of the Census, the population of Mexico was 84,748,000 in 1990 and 93,325,000 (rounded to the nearest thousand) in 1995. Compare these values to your model's prediction from part *e* and explain what happened to the annual growth rate in the years after 1988.

____

____

____

g. If we assume that Mexico's population increases according to the growth factor found in part *b*, then how long will it take the 1982 population to double? *Hint:* Set your formula from part *d* equal to twice the 1982 population and then solve for t by using logarithms. Give an exact answer and then approximate to the nearest whole year.

h. Graph the model you developed in part *d* and label the points representing the initial population and where this initial amount doubles.

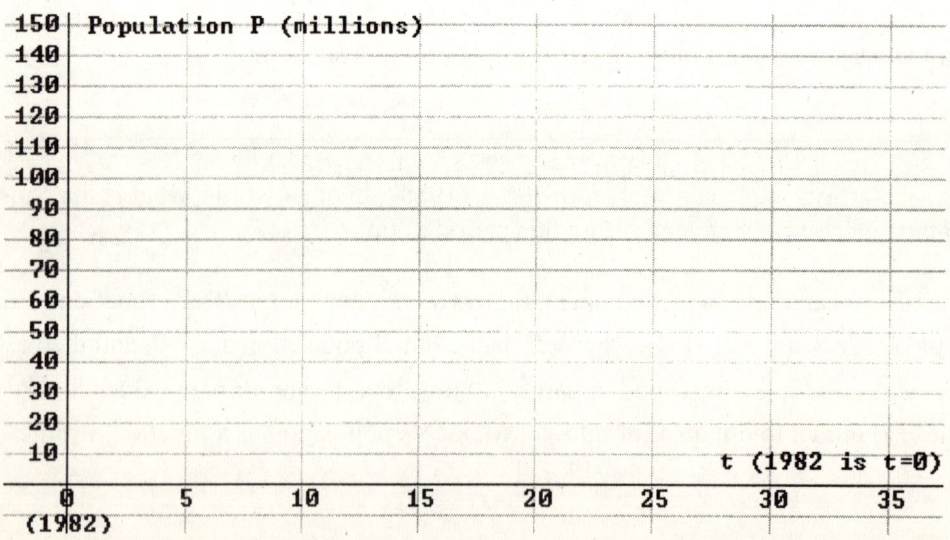

3. Conceptual Exercise—Sweepstakes Winner

Ed has just declared you the winner in the Grand Publisher's Sweepstakes. You have a choice of two prizes:

- Today you receive $150,000 in one lump sum,

OR

- Today you receive a super savings bond called Better Bond that has an initial value of $1 but doubles in value every year until it matures in 20 years. *Hint:* When the money "matures," it can be cashed in.

a. Use the space in the second column of the table to express the doubling as repeated multiplication by 2. Use the third column to express the repeated multiplication in exponential form. For example, tripling the number 1 four consecutive times can be expressed in repeated multiplication form as $1 \cdot 3 \cdot 3 \cdot 3 \cdot 3$ or in exponential form as $1 \cdot 3^4$.

Time (years)	Better Bond Value (dollars)	
	Use Repeated Multiplication	**Use Exponential Form**
0	1	2^0
1	$1 \cdot 2$	$1 \cdot 2^1$
2		
3		
4		
5		

b. Suppose that twenty years have gone by and the bond has matured to its full value.

- How many times have you doubled the initial value of $1?
- How many 2's are multiplied by $1 to obtain the value at maturity?
- Use exponential notation to express the value at maturity.

c. If we let t be the time in years and B be the value of Better Bond, then express this function in terms of the variable t.

$$B(t) =$$

d. How much is Better Bond worth after 20 years?

$$B(20) =$$

e. Suppose that you take the \$150,000 and deposit it in a savings account at Low Interest Bank. If the account has an interest rate of 3% per year, then fill in the following table to see how the balance grows over 5 years. Assume that interest is paid at the end of each year and that interest is added to the previous balance but no additional deposits are made. If necessary, round to the nearest dollar.

Time (years)	Previous Balance (\$)	Interest Calculation	Ending Balance (\$)
0			150,000
1	150,000	$.03 \bullet 150{,}000$	154,500
2			
3			
4			
5			

f. From the table above, note that 100% of \$150,000 plus 3% of \$150,000 is the same as 103% of \$150,000, calculated as $1.03 \bullet 150{,}000$. Each successive year we take 103% of the previous years' balance. Complete the following table to show the necessary calculations using repeated multiplication or using exponential form.

Time t (years)	Low Interest Bank Account Value $L(t)$ (\$)	
	Repeated Multiplication	Exponential Form
0	150,000	$150{,}000(1.03)^0$
1	150,000(1.03)	$150{,}000(1.03)^1$
2	150,000(1.03)(1.03)	$150{,}000(1.03)^2$
3		
4		
5		

g. As time t increases, observe the pattern that is occurring in the exponential form (column 3 of table). Write a function that explains how Low Interest Bank account value L is a function of time t.

$$L(t) =$$

h. Find the value of the Low Interest Bank account 20 years after the initial deposit of \$150,000. Then explain which prize you would prefer. Give reasons for your choice.

$$L(20) =$$

____

____

4. Group Activity—Simple Scales With Logarithms

A quantity known as pH is used as a measure of the acidity or alkalinity of a solution. The pH value is calculated as the negative logarithm of the solution's hydrogen-ion concentration, $[H^+]$, measured in moles per liter. The formula for pH is shown below.

$$\text{pH} = -\log([H^+])$$

Swimming pools and aquariums are good examples of water that must be regularly tested to maintain the proper pH level for a healthy environment. Both are very sensitive to small changes in $[H^+]$ with pools needing to keep their pH at a level that stops microorganisms from growing, and aquariums needing a pH level that allows microorganisms to grow.

a. What is the pH of vinegar with a hydrogen-ion concentration of $[H^+] = 1 \times 10^{-3}$?

b. What is the pH of orange juice with a hydrogen-ion concentration of $[H^+] = 2 \times 10^{-4}$? Round to the tenths place.

c. This table gives the hydrogen-ion concentration $[H^+]$ (in moles per liter) of a number of solutions. Find the pH of each solution, and record the results in the third column.

Solution	$[H^+]$ **(mol/L)**	**pH**
Gastric Juice	1×10^{-1}	
Black Coffee	1×10^{-5}	
Pure Water	1×10^{-7}	
Household Ammonia	1×10^{-11}	

d. The figure below gives the pH scale of acidity and alkalinity. Find the value for $[H^+]$ that corresponds to each pH and express it as a power of 10 below each vertical mark. Note that greater the $[H^+]$, the more acidic the solution.

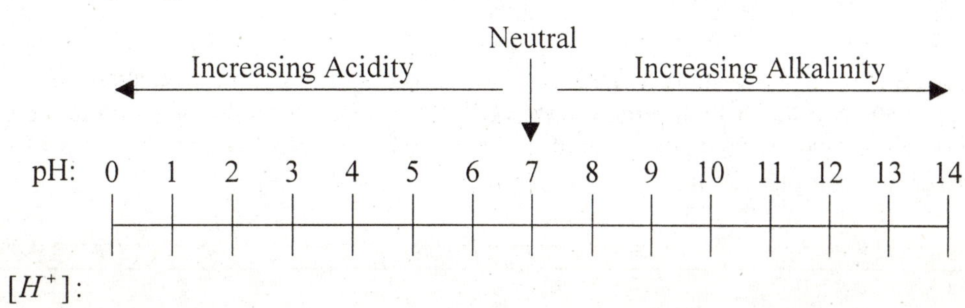

e. Why do you think that pH is measured as the negative logarithm of $[H^+]$ instead of just the hydrogen-ion concentration $[H^+]$ by itself?

____

____

____

f. Complete the following sentence referring to the scale in part *d* as needed.

If pH decreases by one unit, then the hydrogen-ion concentration $[H^+]$ ____________

____

g. How much more acidic is a solution with a pH of 3 than a solution with a pH of 6? Answer in a complete sentence.

____

____

h. The pH of tomato juice is 4.1. Set up and solve a logarithmic equation to find the hydrogen-ion concentration $[H^+]$ of tomato juice.

i. Find the hydrogen-ion concentration $[H^+]$ for the solutions in the table below.

Solution	**pH**	$[H^+]$ **(mol/L)**
Lemon Juice	2.3	
Beer	4.2	
Milk	6.6	
Blood	7.4	

j. Acid rain in some areas of the northeast has a pH of 4.2 while unpolluted rain has a pH of about 5.6. How much more acidic is acid rain compared to unpolluted rain? Show all work and answer in a complete sentence.

____

____

5. Graphing Calculator Activity—The Value of Money

Inflation is the rate at which prices for basic consumer goods and services have increased over time.

a. Complete the following sentence:

As the cost for goods and services increases, the value of money to purchase these same goods and services ______________________________.

b. Suppose that you go shopping today and buy \$100 worth of goods and services. If the annual inflation rate is 5%, then after 1 year, the cost of those same goods and services becomes $\$100 \cdot 1.05 = \$105$. What value (to the nearest cent) of goods and services can be purchased using the original \$100? *Hint:* \$100 is what percent of \$105? Or solve $x \cdot 1.05 = 100$ for x.

c. First we can make a model of how a 5% steady inflation rate affects the cost of \$100 worth of goods and services over time. If we let C represent the cost (in dollars) of goods and services and t represent time (in years), then express C as an exponential function of t.

$$C(t) =$$

d. Find the future cost of \$100 of goods and services and round to the nearest dollar.

Time t (year)	0	1	2	3	4	5	6	7	8	9	10
Cost C (\$)	100										

e. Plot the data from part *d* on the rectangular coordinate system below.

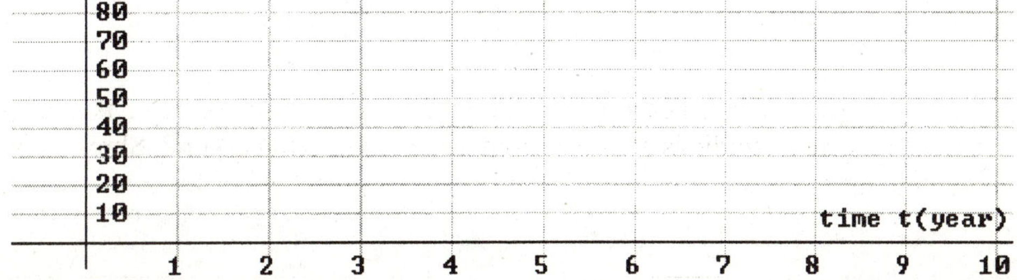

f. Find a function that models the purchasing power of \$100 into the future. Let V be the value of today's \$100 over time t (years). *Hint:* The initial amount of \$100 decays by $\frac{100}{105}$ (about 95.24%) each year. $V(t) =$

g. Use the function above to find the value of \$100 over time (to the nearest dollar).

Time *t* (year)	0	1	2	3	4	5	6	7	8	9	10
Value *V* (\$)	100										

h. Plot the data from the table in part g on the rectangular coordinate system below.

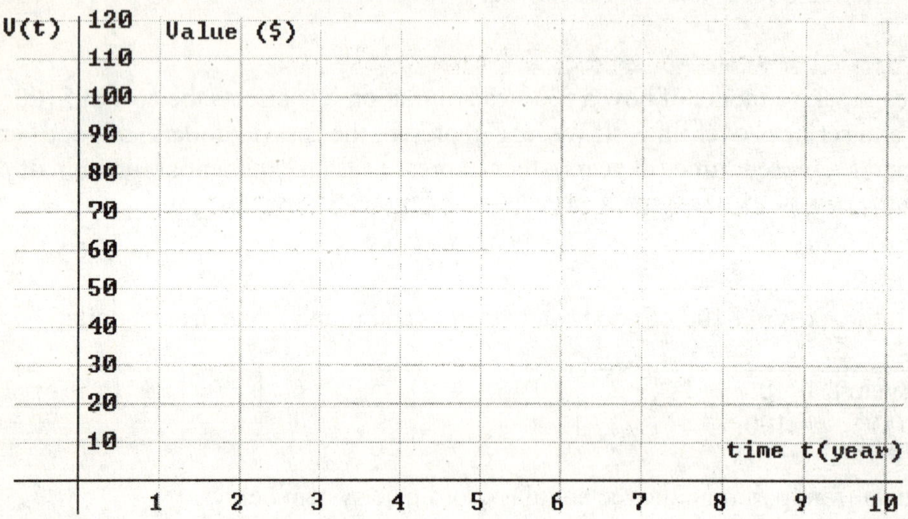

i. Suppose you obtain a \$100,000 life insurance policy today. Assuming a constant annual inflation of 5%, what formula will express the value V over time t in years. If you are 35 years old today ($t = 0$), how much will the policy be worth when you are 70 years old ($t = 35$)?

j. Use the formula from part *i* to find the value to the nearest dollar over time.

Time *t* (year)	0	5	10	15	20	25	30	35
Value *V* (dollars)	100,000							

k. How does inflation effect the value of the policy over time? What things need to be considered before buying any life insurance?

Chapter 11—Sequences, Series, and the Binomial Theorem

1. Extension Exercise—Digital Blues Harp

2. Exploration Activity—Fibonacci and the Golden Ratio in Nature

3. Conceptual Exercise—College Tuition

4. Group Activity—Sharing 2 Things 3 Ways

5. Graphing Calculator Activity—Magic Staircase

1. Extension Exercise—Digital Blues Harp

There are 12 notes in the "A" chromatic scale, A, $A^{\#}$, B, C, $C^{\#}$, D, $D^{\#}$, E, F, $F^{\#}$, G, $G^{\#}$. The next note would be an "A" located exactly one octave above the first A. Starting with the "A" that vibrates at 220 times a second, the (unsimplified) frequencies of the notes of the chromatic scale are shown in the table below.

Note	Frequency (hz)	Note	Frequency (hz)	Note	Frequency (hz)
A	220	$C^{\#}$	$220 * 2^{4/12}$	F	$220 * 2^{8/12}$
$A^{\#}$	$220 * 2^{1/12}$	D	$220 * 2^{5/12}$	$F^{\#}$	$220 * 2^{9/12}$
B	$220 * 2^{2/12}$	$D^{\#}$	$220 * 2^{6/12}$	G	$220 * 2^{10/12}$
C	$220 * 2^{3/12}$	E	$220 * 2^{7/12}$	$G^{\#}$	$220 * 2^{11/12}$

The next note, A, has a frequency of $220 * 2^{12/12}$ = 440 hertz. *Hint:* A chromatic scale can start with any note, then progress through a 12-note sequence in the order listed above. The frequencies are generated with successive terms of the geometric sequence.

In music, adjacent notes in the chromatic scale are "half-steps" apart. The ***diatonic scale*** is a sequence of eight notes from the chromatic scale. The third and forth notes are a half-step apart. The 7th and 8th notes are a half-step apart, and all other adjacent notes are a whole-step (or two-half steps) apart. The 8th note of the diatonic scale is one octave above the first note.

a. In the blanks below, write the missing notes and frequencies that form the diatonic A-scale. Leave the frequencies in unsimplified exponential form.

1st	2nd	3rd	4th	5th	6th	7th	8th
A	______	______	______	______	______	______	A
220 hz	______	______	______	______	______	______	440 hz

b. Any note has a frequency that is double that of the note an octave lower. Find the frequencies of the B and $C^{\#}$ notes of the next octave scale first by doubling the frequencies of the B and $C^{\#}$ from part *a,* then by working through the diatonic scale as you did in part *a,* but start this time with a frequency of 440 Hz for A. Show your work below. Verify that both methods yield the same frequency for each note.

To get the blues sounds digitally, we have to be mechanical and analyze what notes we should add to the scale to get the blues sound.

c. The notes most commonly added to the diatonic scale to get the blues sound are the *flatted third* (1/2-step below the third) and the *flatted seventh* (1/2-step below the 7th note of the diatonic scale. What notes are the flatted third and flatted seventh of the A-scale? What are the associated frequencies of those notes in the scale from part *b*?

flatted third: ______ __________ flatted seventh: ______ __________

note frequency note frequency

d. A major chord is made up of the 1st, 3rd, and 5th note of the related diatonic scale. What notes make up the A-chord in the diatonic scale from part *b*? What are the frequencies of the notes of the A-chord?

1st ______ __________ 3rd ______ __________ 5th ______ __________

note frequency note frequency note frequency

e. To get the blues sound, one has to be able to "slide" from one note to another. It is not unusual to slide from a flatted third to the third of the diatonic scale. As you observed above, the $C^{\#}$-note is the third of the scale and C is the flatted third. Approximate the frequencies of C and $C^{\#}$ (rounded to the nearest frequency). Using those numbers, describe how you might set up an arithmetic sequence of frequencies to slide from a C to $C^{\#}$ over a 1-second interval.

____

____

____

____

____

____

2. Exploration Activity—Fibonacci and the Golden Ratio in Nature

The Fibonacci sequence of numbers starts with $a_1 = 0$, $a_2 = 1$. Any other term of the sequence is sum of the two previous terms.

a. Using the definition above, find the next four terms of the Fibonacci sequence. Show the numbers added to get each term.

$a_3 = a_1 + a_2 =$ ______ + ________ = ____________

$a_4 = a_2 + a_3 =$ ______ + ________ = ____________

$a_5 = a_3 + a_4 =$ ______ + ________ = ____________

$a_6 = a_4 + a_5 =$ ______ + ________ = ____________

b. If two adjacent Fibonacci numbers are 610 and 987, what is the next number? ____________

c. Generate the first 50 Fibonacci numbers and write them in a table such as the one below. What are a_{49} and a_{50}?

$a_{49} =$ ________________________ $a_{50} =$ ________________________

Once the pattern is established, copy down the formulas until the sequence term number in column A is 50.

	A	B
1	Your Name	
2	Class	
3	Fibonacci	
4		
5	Which term:	Fibonacci Number:
6	1	0
7	2	1
8	3	=B6 + B7
9	4	=B7 + B8

d. The number of ancestors of a drone bee forms a sequence with a distinct pattern. A drone is a male and has only one parent, the queen bee. Each female bee (including the queen) has both a male and female parent. Complete the "family tree" of a drone bee up to six generations to verify that the number of ancestors at each level form numbers from the Fibonacci sequence.

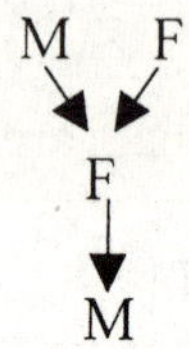

	Total Ancestors
6th level:	____
5th level:	____
4th level:	____
3rd level:	____
2nd level: (grandparents)	2
1st level: (parent is female)	1
Drone Bee is a male (M)	1

e. The Fibonacci sequence appears other places in nature and has a connection to the golden ratio, which is the number, $\frac{1+\sqrt{5}}{2}$.

Find a decimal approximation for the golden ratio rounded to 6 decimal places.

$$\frac{1+\sqrt{5}}{2} \approx \_\_\_\_\_\_\_\_\_\_\_\_\_\_\_\_$$

f. In the Fibonacci sequence, if you divide two adjacent terms (second term divided by the first term), then the ratios get closer and closer to the golden ratio as you go further out in the sequence. Verify that the following ratios are getting closer and closer to the decimal approximation for the golden ratio above. For each computation, write the ratio then compute the decimal approximation.

$\frac{a_5}{a_4} =$ ________ $\approx$ ________ $\frac{a_9}{a_8} =$ ________ $\approx$ ________

$\frac{a_{22}}{a_{21}} =$ ________ $\approx$ ________ $\frac{a_{50}}{a_{49}} =$ ________ $\approx$ ________

3. Conceptual Exercise—College Tuition

Annual tuitions at public and private universities alike are increasing at a rate of 6% per year. During the 1995-1996 academic year, the average tuition was $2860 at a public four-year university and $12,432 at a private four-year university. (*Source: USA Today*, February 20, 1996)

a. In the table below, show what the tuitions would be if the prediction holds true. *Hint:* To compute a 6% increase, add the tuition for the year to 6% of the tuition to get the next year's tuition.

School Year	Public 4-year University	Private 4-year University
1995-1996	$2,860	$12,432
1996-1997		
1997-1998		
1998-1999		
1999-2000		
2000-2001		

b. Find the general term of the sequence that describes the pattern of average annual tuition for public universities. Let $n = 1$, represent the 1995-1996 academic school year.

$a_n =$ ______________________________

c. Find the general term of the sequence that describes the pattern of average annual tuition for private universities. Let $n = 1$, represent the 1995-1996 academic school year.

$a_n =$ ______________________________

d. Assuming that the rate of tuition increase remains the same, find the annual tuition for the 2010-2011 academic year for each type of university.

Public University: ____________________ Private University __________________________

e. Use partial sums to find the average total cost of a four-year college education at a public university for a student starting in the 2000-2001 academic year. Show how you set up your arithmetic expressions for calculations. (Use a calculator for the actual calculations.)

f. Use partial sums to find the average total cost of a four-year college education at a private university for a student starting in the 2000-2001 academic year. Show how you set up your arithmetic expressions for calculations. (Use a calculator for the actual calculations.)

g. Find the year in which tuition at a public university reaches or exceeds the 1995-1996 average cost of attending a private university.

h. Explain why the above costs are only estimates. What other factors might affect the future costs of a student's tuition?

4. Group Activity—Sharing 2 Things 3 Ways

The ideal group size for this activity is 3 members. Materials needed per group: 2 note cards and a pair of scissors.

a. Take one note card and decide how to divide it evenly among the three members of the group. (If there are more than three members, select three group members to get a share of the cards.) How much of the card does each of the three members get?

b. The remaining note card will be distributed among three group members. This time, however, the only allowable cut of a card or piece of card will be 2 equal pieces. Each time cuts are made, distribute as many equal pieces as possible. Then cut remaining pieces for the next distribution. Record how much of a note card is distributed to each of the three people at each step. Make at least two distributions, and then generalize the pattern to predict the next distributions.

Step	Fraction of Original Card Distributed	Size of Piece Remaining	Step	Size of Piece Remaining	Fraction of Original Card Distributed
1			4		
2			5		
3			6		

c. Write the first five terms of the sequence shown in the "Fraction of Original Card Distributed" columns in the table above, and write the *nth* term.

a_1 = ___________ a_2 = ___________ a_3 = ___________

a_4 = ___________ a_5 = ___________ a_6 = ___________

d. Determine the values of a_1 and r to represent the above sequence as a geometric sequence. Check your choices by showing that $a_2 = a_1r$, $a_3 = a_1r^2$, and so forth, gives the same values for a_2, a_3, a_4, a_5, and a_6 as in part *c*.

a_1 = ______________________ r = ______________________

e. Use the formula, $S_n = \dfrac{a_1(1-r^n)}{1-r}$ to find the sum of the first six terms of a geometric series. Show your substitutions. Simplify your answer to exact fraction form, then use your calculator to put your answer in decimal form. Compare your answer to the sum of the first six terms in part *c*.

sum = ____________________ = ____________________

simplified fraction decimal form

f. Substitute the values for a_1 and r into $S(n) = \dfrac{a_1(1-r^n)}{1-r}$. Then use a graphing calculator to graph *S(n)*. Use the variable *x* instead of *n*. Use your "trace" or "calc" feature to estimate *S(30)*. Does there appear to be much difference between *S(30)* and *S(200)*? Explain.

g. Use the formula, $S_\infty = \dfrac{a_1}{1-r}$ to compute the sum of the infinite sequence formed by continuing to divvy out pieces of the note card infinitely. Leave your answer as a simplified fraction. In light of your answer to part *a* and the graph in part *f*, explain how you could have predicted the sum of the infinite series.

$S_\infty =$ ____________________

5. Graphing Calculator Activity—Magic Staircase

Suppose a magic staircase appears before you in the following manner. First, a single step is made from a single block. As soon as you step on the first step, a second step appears. The second step is three blocks high. As soon as you step on the second step, a third step appears (five blocks high) and a fourth step (seven blocks high). As you move up the staircase, the pattern continues. See the diagram below.

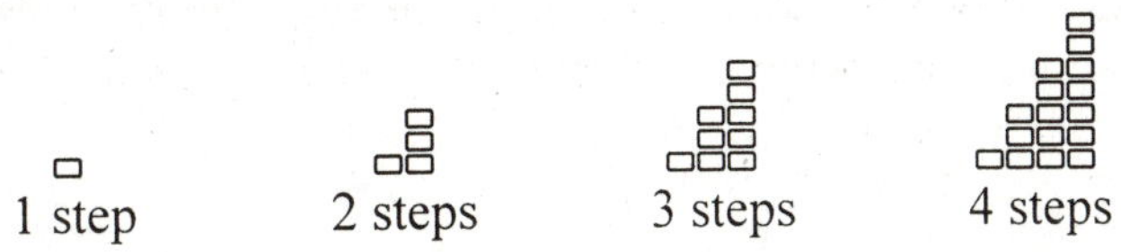

a. Following the established pattern above, draw a 5-step staircase and a 6-step staircase.

b. Fill in the table below to show how many blocks are stacked to make each step and how many total blocks are used to build the staircase. In the last row, generalize your observations to state what happens in the *n*th step. *Hint:* To see the pattern, use the last column to show the numbers as the sum of blocks in each step.

Which Step	**Number of Blocks Stacked to Make the Step**	**Total Number of Blocks in the Entire Staircase up to This Step**
1		
2		
3		
4		
5		
6		
n		

c. The numbers in column 2 of the table in part *b* form an arithmetic sequence. What is the first term a_1, the difference *d*, and the *n*th term of the sequence?

$a_1 =$ ____________ $d =$ ____________ $a_n =$ ____________

d. Write the formula *S(n)* for the sum of the first *n* terms of the arithmetic sequence formed by the blocks and steps. Replace a_1 with its value from part *c* and a_n with its expression from part *c*. Then simplify. Describe what *n* represents and what *S(n)* represents in the magic staircase.

$S(n) =$

n represents ______________________________

S(n) represents ______________________________

e. Use the formula from part *d* to compute the partial sums. In the third column of the table below, write ordered pairs as directed. Compare your answers to the totals in part *a*.

Number of Terms in Partial Sum	Partial Sum	Ordered Pair (Col 1 , Col 2)
1		
2		
3		
4		
5		
6		

f. On your graphing calculator, plot the ordered pairs from the table in part *e*. Graph the curve of *S(n)* from part *d* to confirm that the ordered pairs are on the curve. Does the entire graph of *S(n)* describe the sequence formed by the partial sums in part *e*? Explain your reasoning.

____

____

____

PHim2
Prentice Hall Interactive Math

Student User Manual

K. Elayn Martin-Gay
University of New Orleans

• Includes Installation Instructions •

Prentice Hall

One Lake Street
Upper Saddle River,
NJ 07458

Project Manager, Developmental Math: *Elaine Page*
Project Manager, Media: *Audra Walsh*
Production Editor: *Donna Crilly*
Managing Editor: *John Matthews*
Media Buyer: *Jill Arcade*
Print Buyer: *Michael Bell*
Editor in Chief: *Christine B. Hoag*
Editorial Assistant: *Christina Simoneau*
Senior Marketing Manager: *Eilish Collins Main*
Editor in Chief, Development: *Carol Trueheart*
Senior Design Manager, Software, Print: *Greg Lamb*
Senior Production Manager: *Peter Silvia*
Product Manager, Mathematics: *Ian Seekell*
Associate Product Manager: *Jennifer Mangan*
National Program Consultant, Developmental Math: *Diane Gray*
National Sales Director: *Sean Rooney*
Executive Producer, Prentice Hall Media Group: *Phil Lanza*
Director, Prentice Hall Media Group: *Dave Roh*

Prentice Hall

Upper Saddle River, New Jersey 07458

The author and publisher of this book have used their best efforts in preparing this book. These efforts include the development, research, and testing of the theories and programs to determine their effectiveness. The author and publisher make no warranty of any kind, expressed or implied, with regard to these programs or the documentation contained in this book. The author and publisher shall not be liable in any event for incidental or consequential damages in connection with, or arising out of, the furnishing, performance, or use of these programs.

Printed in the United States of America

10 9 8 7 6 5 4 3 2

ISBN:0-13-102018-8

Pearson Education Ltd., London
Pearson Education Australia Pty. Ltd., Sydney
Pearson Education Singapore, Pte. Ltd.
Pearson Education North Asia Ltd., Hong Kong
Pearson Education Canada, Inc., Toronto
Pearson Educacíon de Mexico, S.A. de C.V.
Pearson Education—Japan, Tokyo
Pearson Education Malaysia, Pte. Ltd.
Pearson Education, Upper Saddle River, New Jersey

GETTING STARTED

Welcome to Prentice Hall Interactive Math 2, an on-line, computer-based course for math students.

Here are just some of the ways that you will learn math using this program:

- Watch videos at the objective or section levels,
- Read the Review, Reference, and Practice textbook in print or on-line in Read e-book,
- Work through problems at your own pace using Step by Step,
- View an animated problem-solving using Demonstrate,
- Solve Practice and Assess exercises for a specific objective,
- Take diagnostic tests to find out how much math you already know and which skills require more study and practice.

For the most up-to-date information on Prentice Hall Interactive Math 2, please visit the following website: http://www.phinteractivemath2.com/support.

SYSTEM REQUIREMENTS

Before using the Prentice Hall Interactive Math 2, please make sure that your computer meets the following minimum system requirements:

CPU	Pentium II 300 MHZ
RAM	128 MB
Video	SVGA
Resolution	800 x 600, Thousands of colors (16 BIT)
Sound Card	Sound Blaster or Sound Blaster-compatible
Network Card	Any network card. Not required if dialing from off-campus through an Internet connection.
Monitor	15" monitor
CD-ROM Drive	16x
Disk Space	200 MB (100 MB if Internet Explorer is already installed)
Internet Connectivity	Campus Network or 56Kbps Modem with a certified Internet Service Provider
Operating System	Windows 98, Windows NT, Windows ME, Windows 2000, Windows XP
Internet Browser	Microsoft Internet Explorer 5.5 or higher (Internet Explorer 6 provided)

Plug-ins Provided on the Prentice Hall Interactive Math 2 CD-ROM:

- QuickTime 5 or higher (Link to QuickTime Website for installation of QuickTime 6)
- Java 1.3.1 Plug-in
- TestGen Plug-in
- Flash 6 Player (Link to Macromedia Website for installation)
- Shockwave 6 Player (Link to Macromedia Website for installation)

INSTALLATION INSTRUCTIONS

Please use your Prentice Hall Interactive Math 2 CD to install the software plug-ins necessary to access the program. CD installation is **recommended** but not required since all plug-ins can also be downloaded from the Prentice Hall Interactive Math 2 support site. Depending on your Internet connection speed, it is usually quicker to install the required plug-ins from the CD than to download them from the Internet.

To install the components needed to successfully run Prentice Hall Interactive Math 2:

1. Insert the CD in the CD-ROM drive.
2. Go to the **Start** menu; select **Run**.
3. Type **x:setup.exe** (where **x** is the letter of your CD-ROM drive).
4. Follow the instructions on screen to complete the installation.

Setup will check your system for the presence of the necessary software to run Prentice Hall Interactive Math 2, including Internet Explorer 5 or higher, Sun Java plug-in, QuickTime 5 or higher, and the TestGen plug-in. If any of these components are missing or are not in their latest versions, the Setup program will offer to install them for you.

Once the Setup has installed the items you need, a web page will open with a link to the Prentice Hall Interactive Math 2 support page. The support page contains links that will verify that you have the latest versions of QuickTime, Flash, and Shockwave. All three of these items are required and should be installed before you use the program. If you do not have the current versions, they should be installed from their respective websites by clicking the links provided and following the instructions.

Note: Even if you already have Internet Explorer installed as your browser, it is recommended that you install from the program CD to ensure that all needed components are present. Your present settings for Internet Explorer will not be disturbed in the process.

If you need technical support, please contact:

Prentice Hall Media Support Line:

1-888-695-6577, Option 3

Hours of Operation: Monday –Thursday 9:00 AM – 10:00 PM
Friday 9:00 AM – 5:00 PM, and
Sunday 5:00 PM – 10:00 PM EST

Email: media.support@pearsoned.com

STUDENT REGISTRATION

1. To access Prentice Hall Interactive Math 2, go to the Internet and use Internet Explorer as your browser.

2. Enter the URL: http://www.phinteractivemath2.com

3. If this is the first time you have accessed Prentice Hall Interactive Math 2, you will be asked to run the Readiness Wizard. If you have already installed the plug-ins from the CD-ROM, you may want to skip this step. However, run the Readiness Wizard in order to ensure the most up-to-date plug-ins.

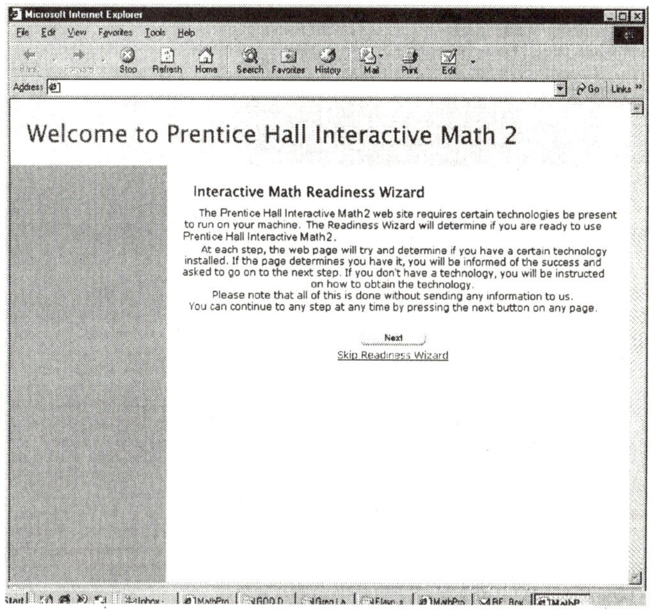

4. To use the Readiness Wizard, begin by clicking the **Next** button. You will run through a short series of screens that check for the necessary plug-ins and ask if you would like to download and install them. If everything is up-to-date, click the **Next** button at the bottom of the page. If you need to download a plug-in, follow the on-screen instructions for downloading and installation.

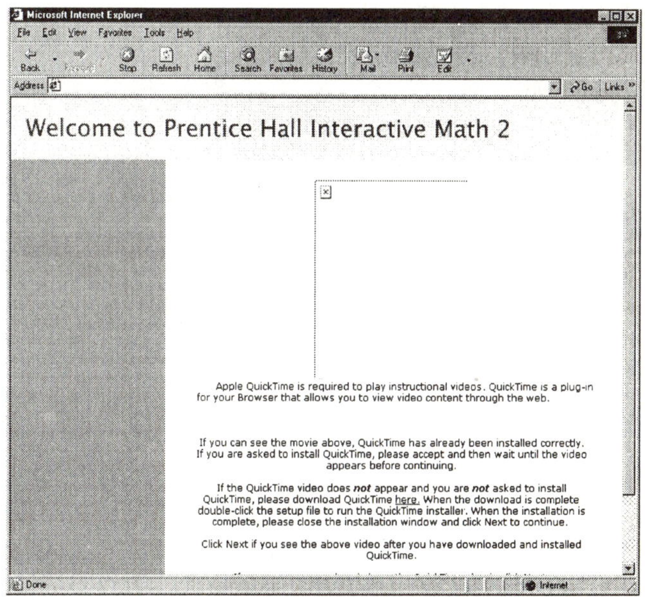

5. Once the Readiness Wizard is complete, the Welcome page for Prentice Hall Interactive Math 2 will open. Select the **Click Here** button to open the Login screen.

Note: The Prentice Hall Interactive Math 2 Welcome page will open automatically on subsequent visits to the site. You will not need to press **Click Here** to launch the program every time.

6. After pressing **Click Here**, you will arrive to the **Login** screen. To register as a new Prentice Hall Interactive Math 2 user, click on the **New User** button.

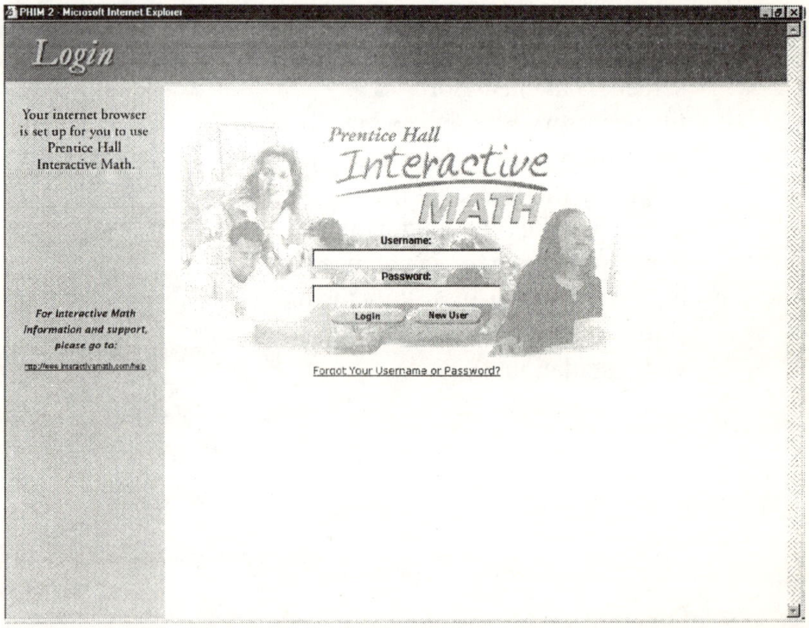

7. After clicking **New User**, you will see the Registration screen with **1. Your Access Code** listed at the top. The blue panel on the right shows your progress through the registration steps.

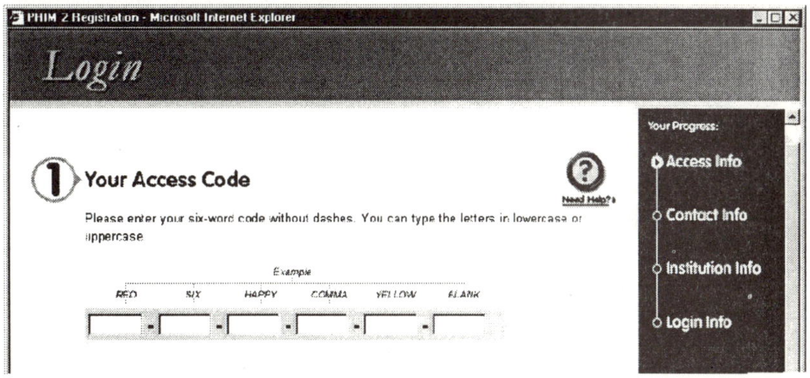

8. In the rectangular boxes shown for **Your Access Code**, enter the 6-word access code found on the inside, pull down tab of your Prentice Hall Interactive Math 2 CD package.

9. Click the **Next** button to continue with the registration.

10. The next screen is **Your Contact Information**. Enter your name, phone number, email address, and your college's zipcode in the appropriate boxes.

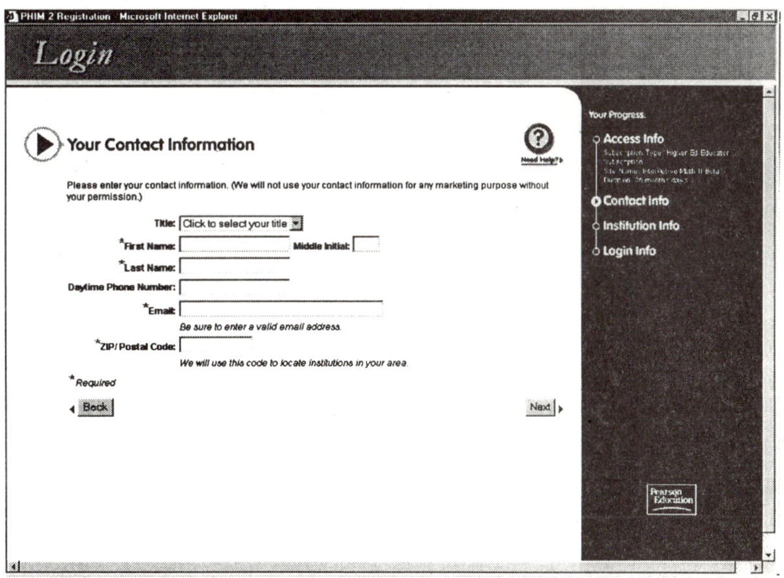

Note: Be sure to enter the zip code for your *college* as this is used to locate colleges in your area.

11. Click the **Next** button to go to the next page. Notice that the blue panel on the right keeps track of the information you have entered in previous screens.

12. This screen is **Your Academic Institution**. Use the drop down box labeled **Institution Name** to select the name of your college. If your college is not listed, select "Other" at the bottom of the list.

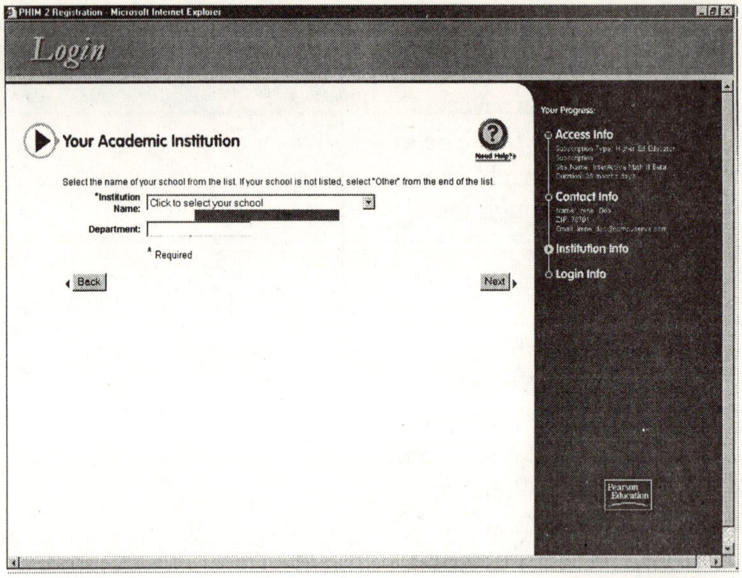

Note: If you selected "Other" from the list of **Institution Names**, you will see an additional screen (shown at right) where you enter the information for your college. Then enter the name of your college next to the box for **Institution Other,** and enter the city, state, and country in the appropriate boxes.

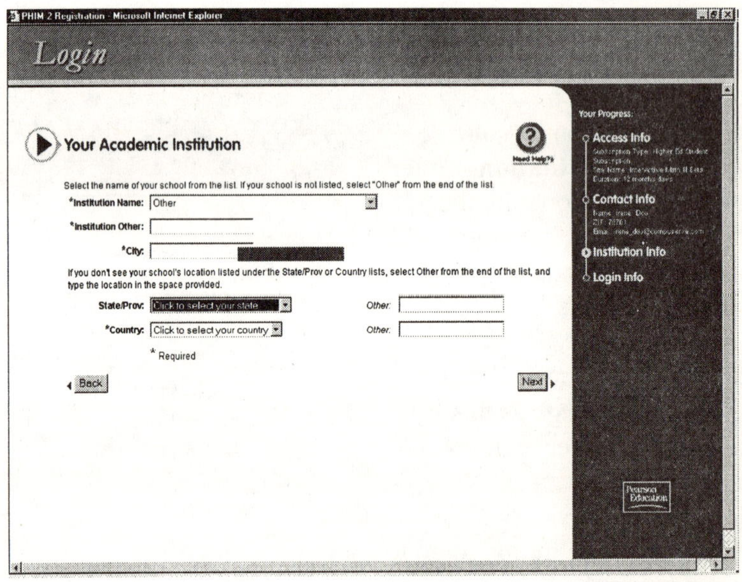

13. Click the **Next** button to continue.

14. The next screen starts with **Your Login Name and Password**. Make your login name unique, such as *first name + last name*. If you have a common name such as John Smith, add numbers to the end of the name, **johnsmith1234**. Do not use spaces.

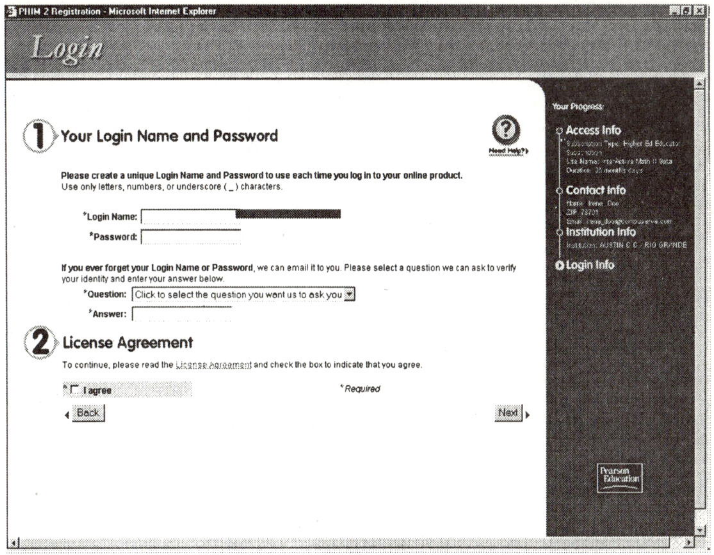

15. Select a question to use for verification in the event that you forget your name or password. Then type in the answer.

16. The next part of the screen is called **License Agreement**. Check the "I agree" box and then click the **Next** button to complete your registration.

17. The **Confirmation and Summary** screen shows the results of your registration. An email with the same information will be sent to the email address you supplied earlier. Keep this email in a safe place in case you need to retrieve your login name and password.

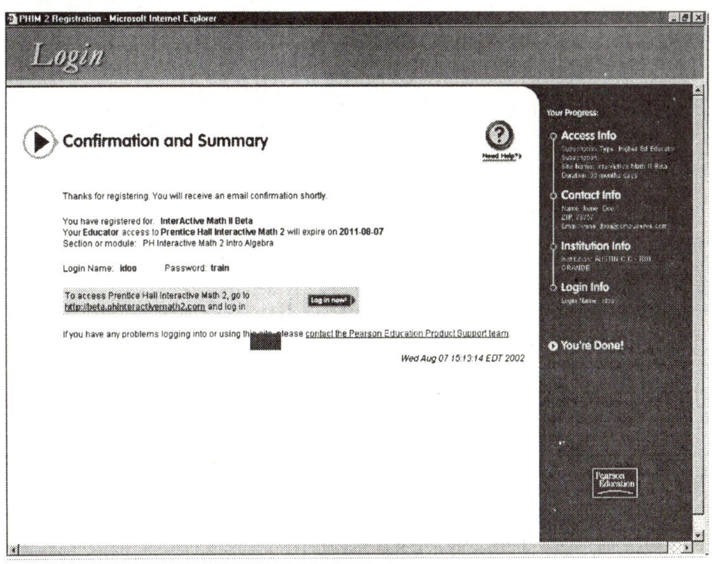

18. Close the browser window when you are done, *or* click the **Login Now** button to open the program.

STUDENT LOGIN

After registration is complete, the **Login** screen will appear again.

1. Enter your **Username** and **Password**. Then click the **Login** button.

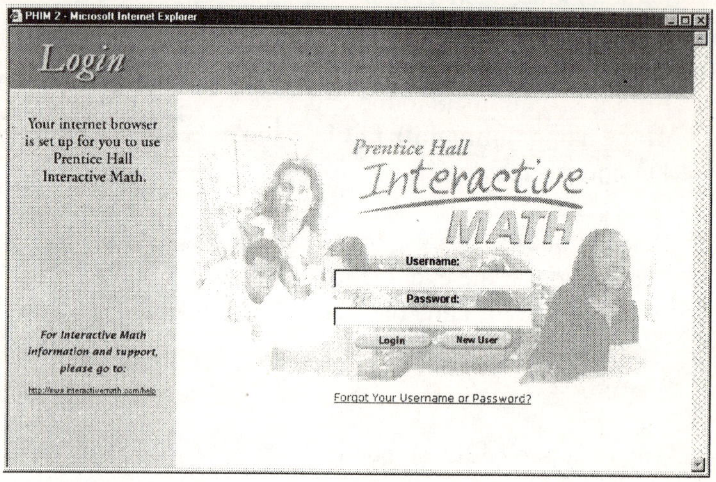

2. Enter the **Course Code** provided by your instructor and click on the **Submit** button. Contact your instructor if you do not have the Course Code.

Note: If you click the "**Click here to start over**" link, you will return to the **Login** screen.

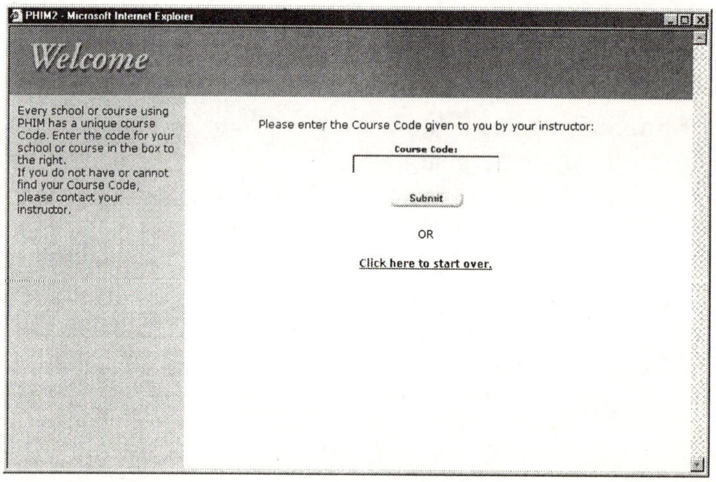

After you have entered in the **Course Code**, you will see a confirmation screen.

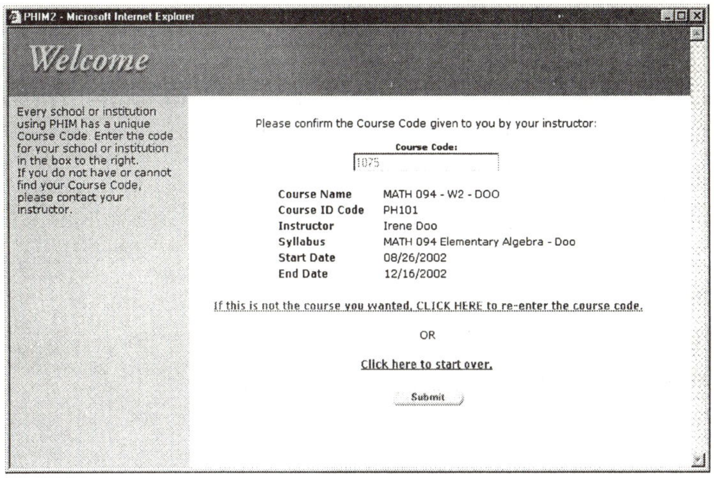

3. Check that you have selected the correct **Course Name** and **Instructor**. If the information is correct, click the **Submit** button to enter the course.

Note: If you have selected an incorrect course, click the first link to re-enter your **Course Code**. If you need to start the **Login** process again, click the second link, which reads as "Click here to start over."

HOW TO USE PRENTICE HALL INTERACTIVE MATH 2

When you first login with Prentice Hall Interactive Math 2, you will see a screen like the one at right, which shows the course(s) for which you are registered.

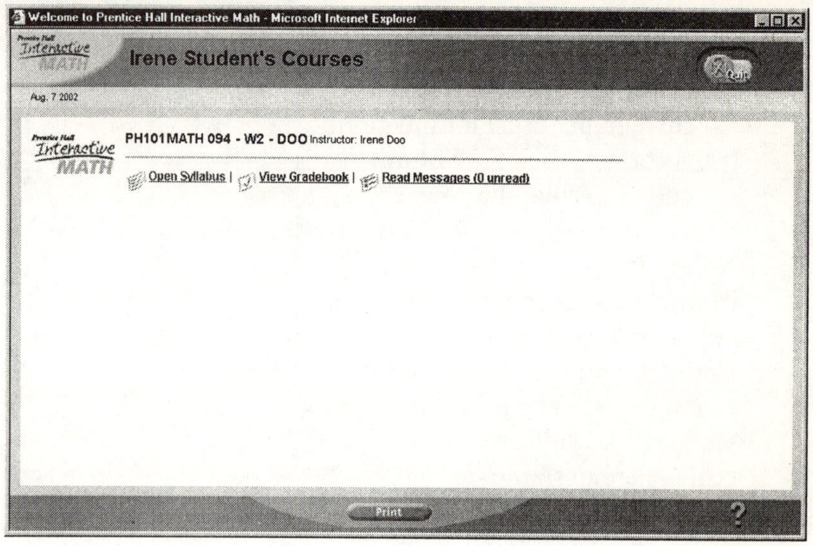

You will see the **Open Syllabus**, **View Gradebook**, and **Read Messages** icons.

The **Print** button allows you to print out this screen. The **?** button at the bottom right will provide software help. To exit the program, click the **Quit** button at the top right corner.

Now click the **Open Syllabus** icon to enter your course.

SYLLABUS

The **Syllabus** screen gives you a map for the entire term, as shown to the right. Depending on how your instructor has designed the syllabus, you may see a syllabus with weeks and dates or a chapter-by-chapter syllabus.

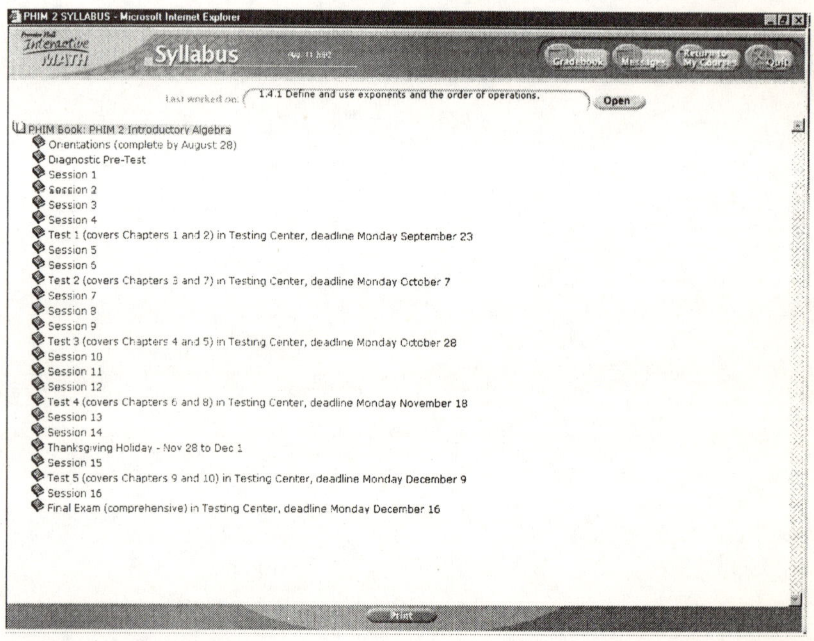

In the **Syllabus**, you can open a syllabus item by clicking on the icon to the left of the item name.

Note: If you have previously worked in the program, the last syllabus item you were working on will be displayed in the **Last worked on** box. Click the **Open** button to the right of this box to go directly to this syllabus item. In the example on the right, clicking the **Open** button will take you directly to objective 1.3.1.

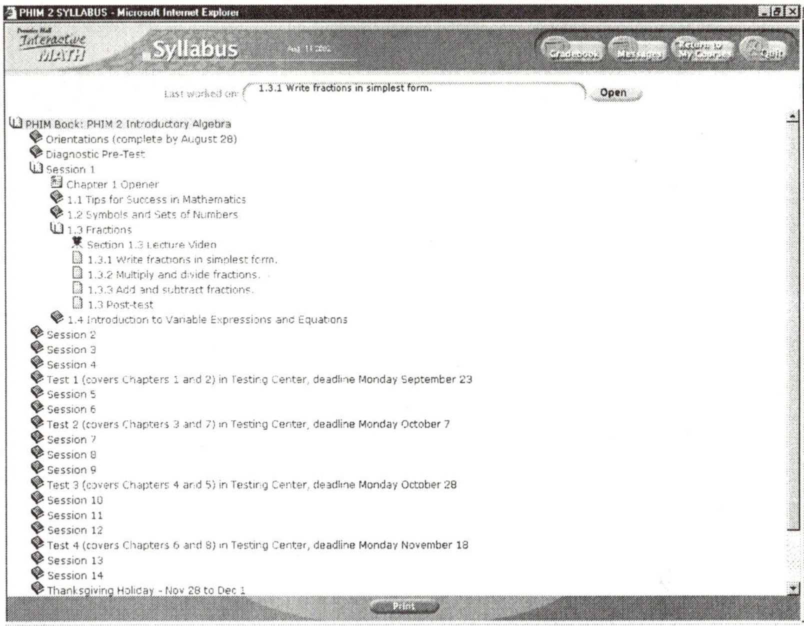

Items that have a book icon may contain additional items inside. In the screen on the right, the **Session 1** book is open, and the **1.3 Fractions** book is inside and open to reveal the individual syllabus items within that section.

The different syllabus icons are described on page 24 of this manual.

TESTING

Depending on how your instructor has created your course syllabus, you may take several types of tests:

- Diagnostic Test
- Section Test
- Chapter Test, or
- TestGen Test

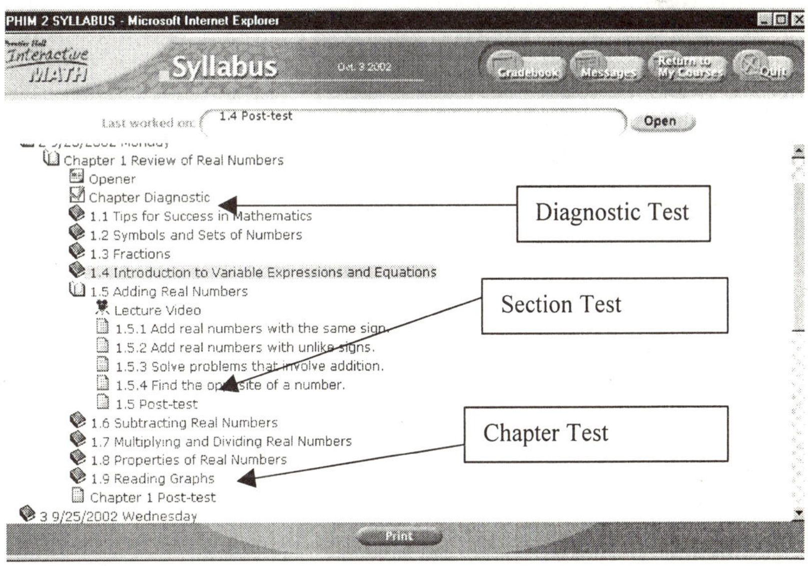

DIAGNOSTIC TESTS

Diagnostic Tests are designed to measure your current skills—what you know and don't know about math. Your instructor may use the results to help you gain any necessary prerequisite knowledge or to allow you to skip some objectives in the syllabus.

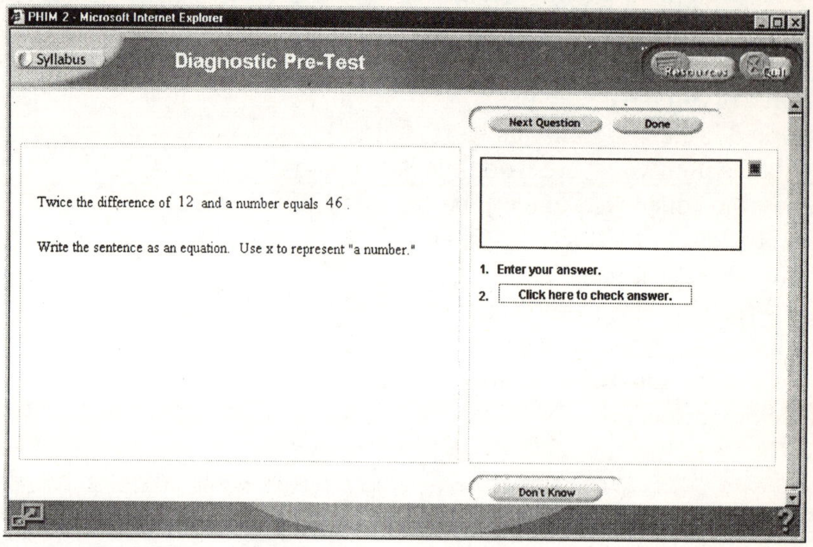

Type your answer in the answer box to the right of the question. Then press the **Click here to check answer** button to submit your answer.

If you do not know how to answer the question, click the **Don't Know** button.

Click the **Next Question** button to go to the next problem in the test.

Click the **Done** button to stop the test.

When you click **Done**, you will be given three choices, shown at right. Select the appropriate response and click the **Submit** button.

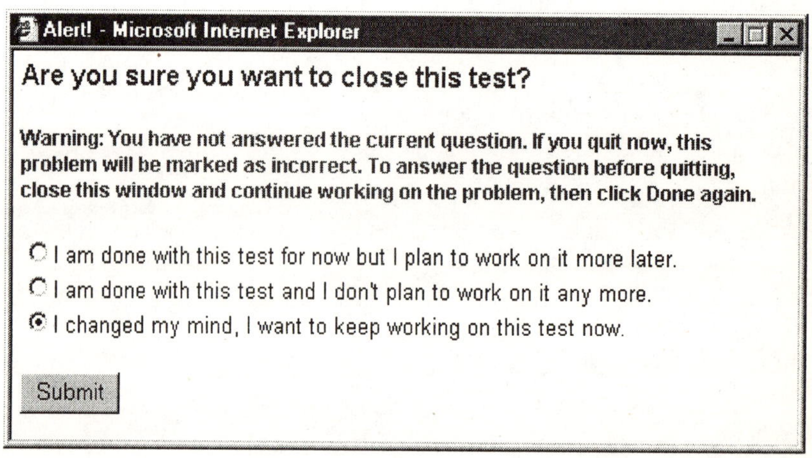

Note: Depending on how your instructor set up the diagnostic test, some of these choices may not be available.

Depending on your choice, the next screen may be similar to the one on the right.

Click on **Click here** to return to the **Syllabus** screen.

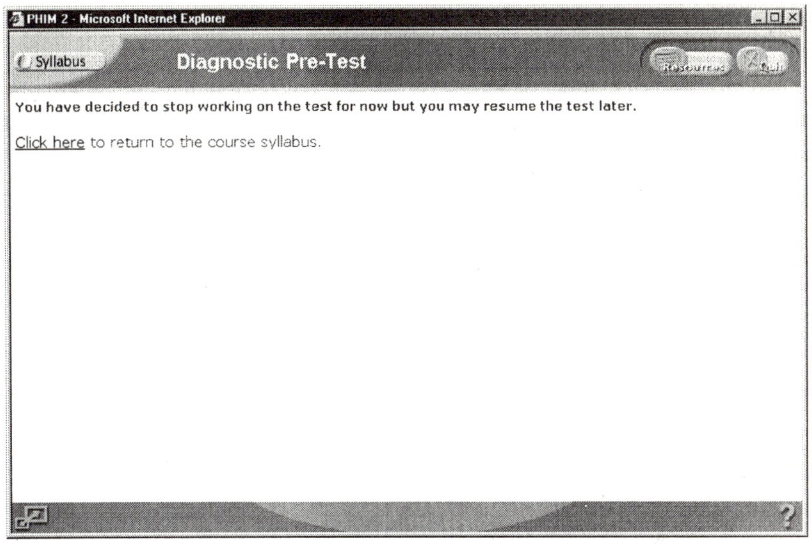

SECTION TESTS

On your course syllabus, you may see tests that appear at the beginning of a section, at the very end of a section, or both.

The screen on the right shows an example of a **Section Post-Test.**

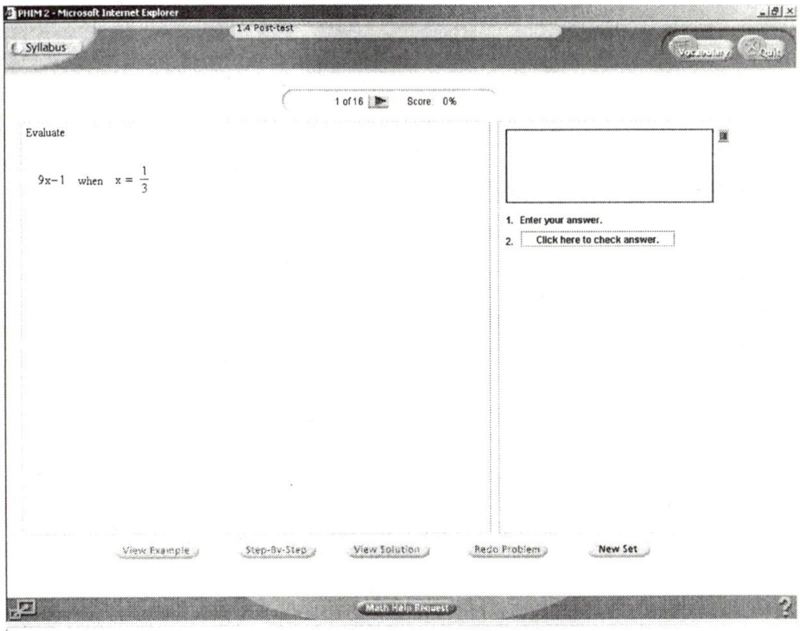

CHAPTER TESTS

On your Syllabus, you may see tests that appear at the beginning of a chapter, at the end of a chapter, or both. These chapter-level tests are similar in format to the **Section Tests.**

The screen on the right shows an example of a **Chapter Post-Test.**

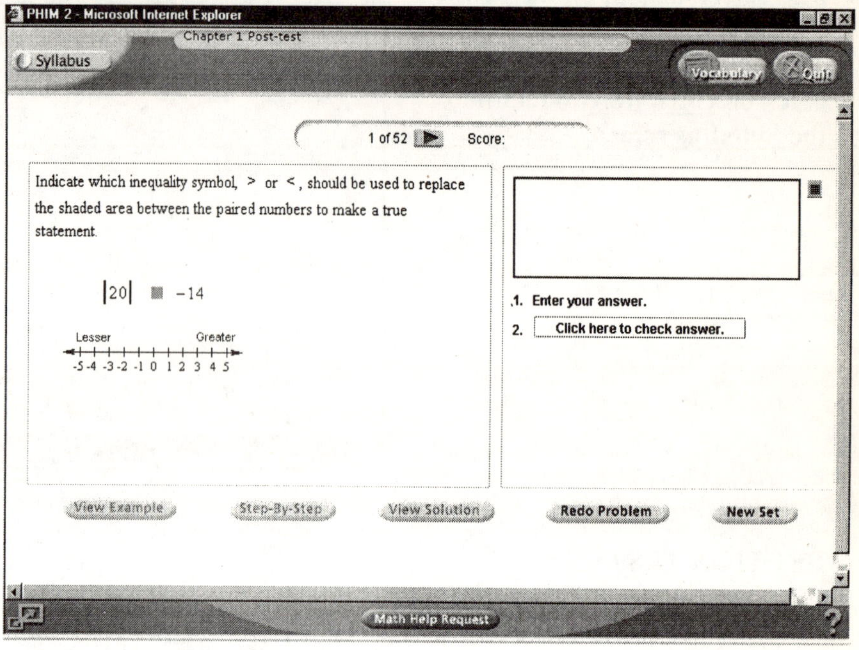

ONLINE TESTS (TestGen Tests)

Your instructor may choose to assign additional online tests such as the TestGen test, shown at right.

Each question appears on its own screen. When the test is complete, you can check to see both correct and incorrect answers for a comprehensive review of your progress.

The TestGen plug-in required to take these tests online is provided during the Prentice Hall Interactive Math 2 installation process.

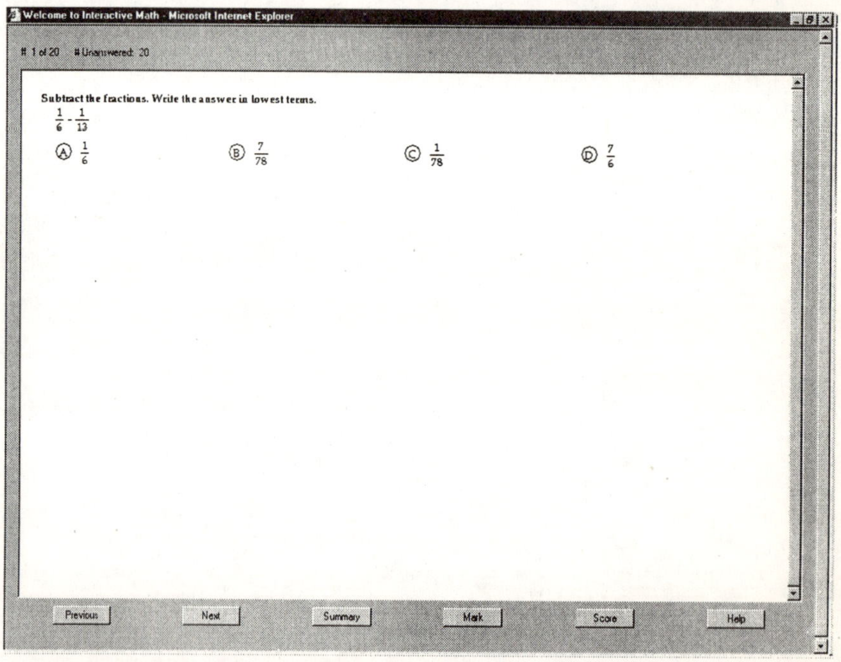

WAYS TO LEARN MATH

The **Chapter Openers** are fun activities that introduce you to a main concept in that chapter.

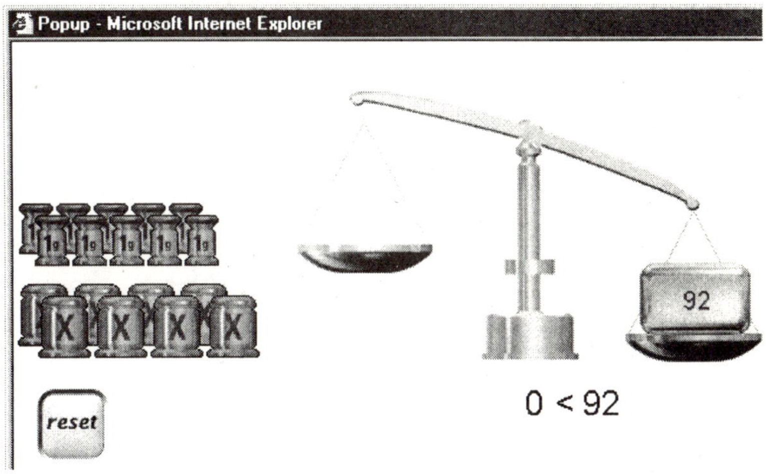

Lecture Videos appear at the section level and show author K. Elayn Martin-Gay walking through specific problems for that particular section. You can pause or replay the video as often as you like.

Lecture videos stream off the Internet.

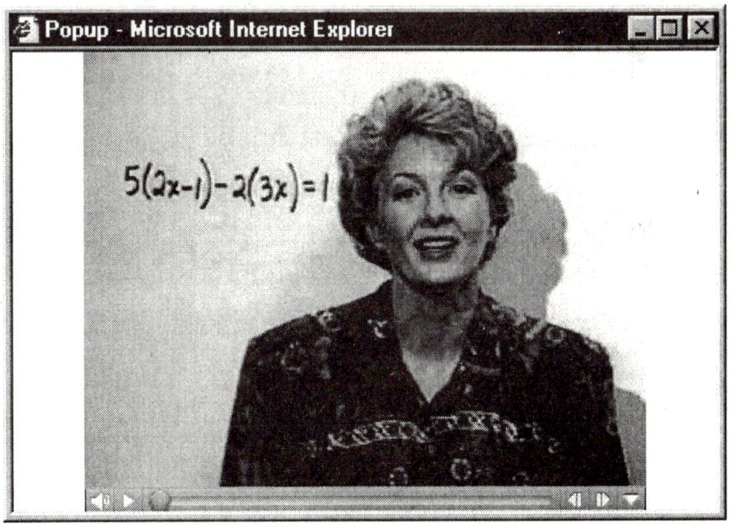

Each objective has six main screens – **Intro, Read e-Book, Watch, Step by Step, Practice**, and **Assess**.

Each screen shows you the concept in a different approach. You can work through an objective in any order you choose.

The **Intro** screen gives you a summary of the mathematical concept.

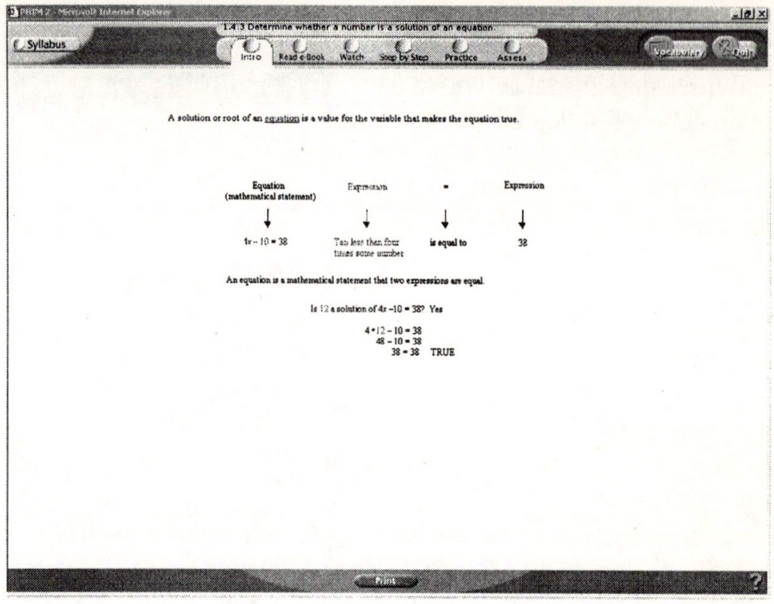

Read e-Book shows you a written lesson and examples from the objective. Scroll down to read the entire lesson. Each Read e-Book screen corresponds to the same numbered objective in your accompanying **Review, Reference, and Practice** textbook.

Click any blue **hyperlinks** to see an explanation of the word or phrase.

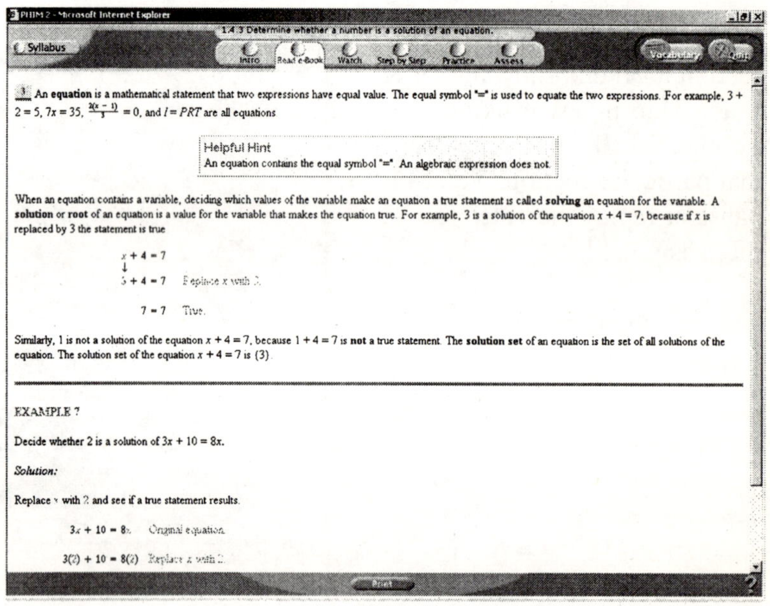

The **Watch** screen provides a short video (about 1-2 minutes) of author K. Elayn Martin-Gay working through a problem on a whiteboard. You can pause or replay the video as often as you like.

You can steam these videos off the Internet or use the program CD.

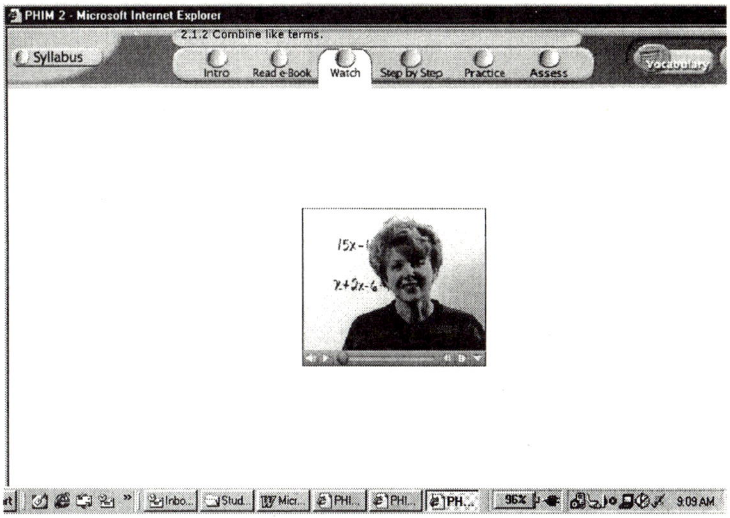

Step by Step guides you through the problem-solving process in stages.

Read the problem and enter your answer in the answer box on the right side of the screen. Click where it says to "Click here to check answer". Continue working through the problem and answering questions until the problem is solved.

For more practice, click **Work Another Problem**.

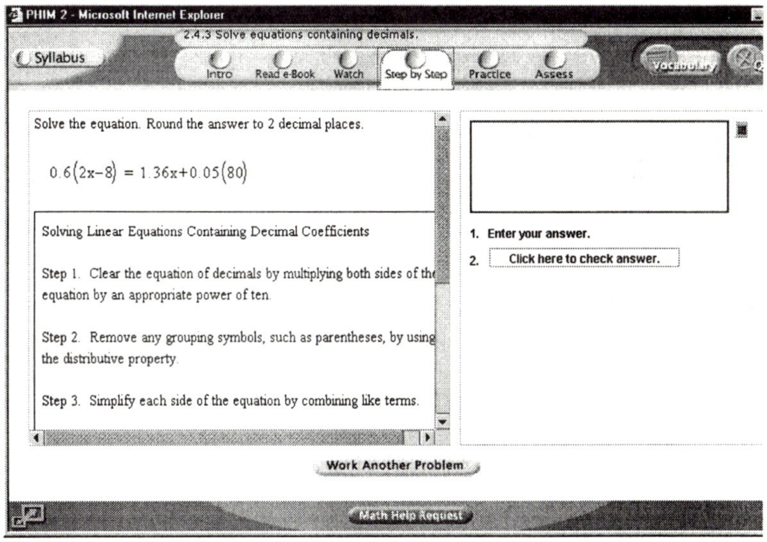

The **Practice** exercises give you a set of problems to practice your skills from this objective. These problems will not count toward your grade. Depending on how your instructor set up your syllabus, some buttons may not be available.

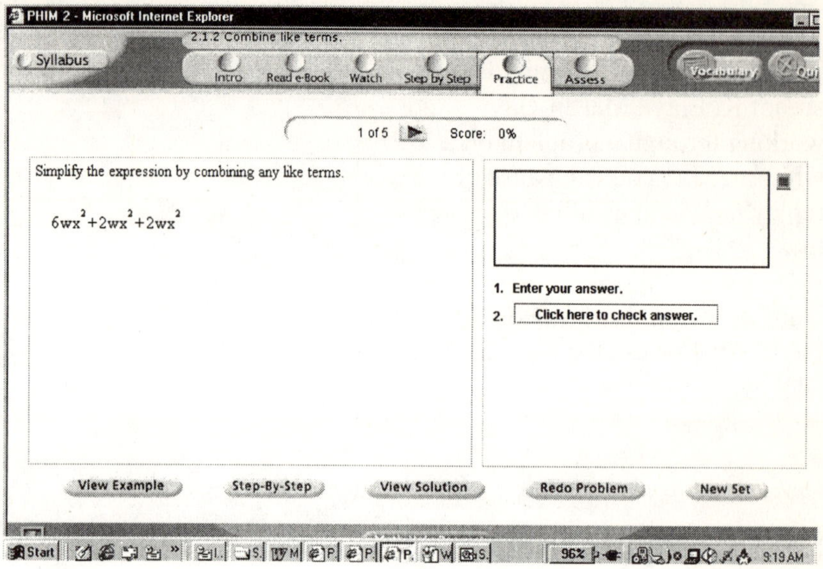

Read the problem and enter your answer in the answer box on the right side of the screen. Click the **Click here to enter your answer** button or press the **Enter** key to submit your answer.

You can also resize this screen by using the **Resize Math Area** button located in the bottom left corner. Click your mouse at this corner and drag the screen to the desired size for viewing.

Click **View Example** to see a different problem and its solution. **Step-By-Step** allows you to work through the problem in stages.

Click **View Solution** to see the solution to the given problem. **Redo Problem** generates a different problem to work.

Click **New Set** to generate a new set of problems to work.

The **Assess** screen is very similar to the Practice screen, and the options available depend on how your instructor has set up the syllabus. In the example shown at right, only the **Redo Problem** and **New Set** buttons are available.

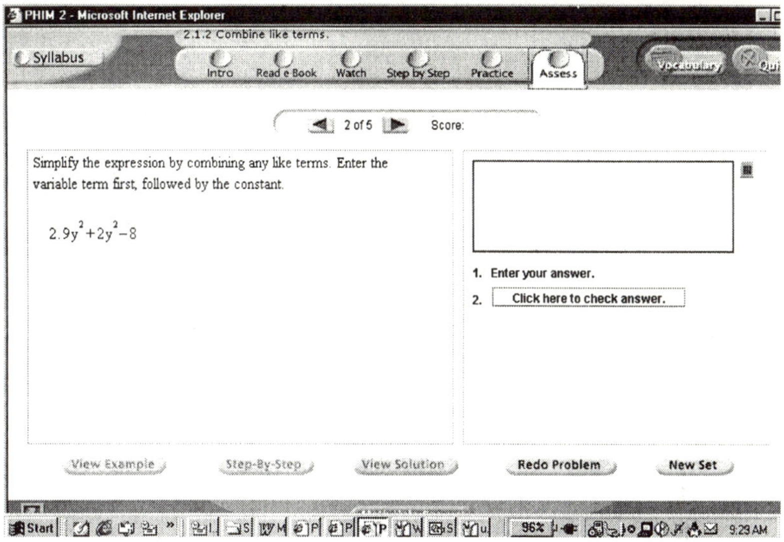

Like the Practice screen, you can send a **Math Help Request** and resize this window. Always return to the **Syllabus** when finished with an objective so that your work will not be lost. Do not close your Internet browser window or your grades will not be counted.

Demonstrate appears for specific objectives that instructors have identified as especially difficult for their students.

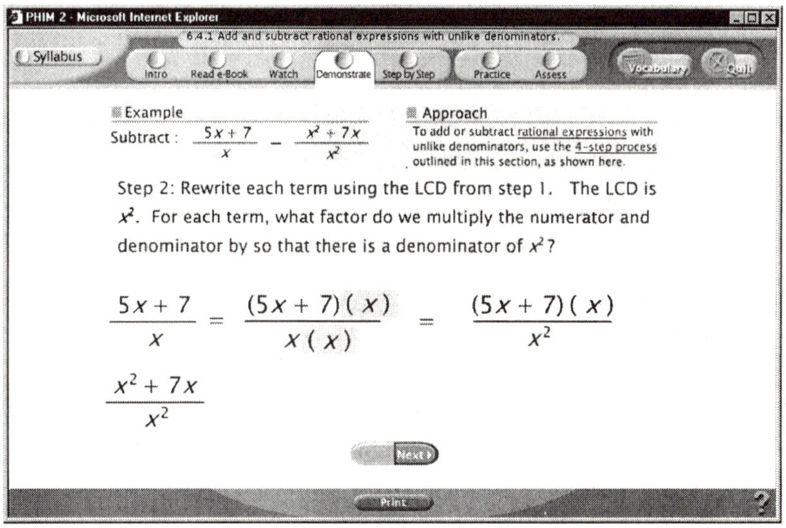

Using animation, audio, visuals, and text, **Demonstrate** shows you how to solve a particular example in the objective.

When prompted, click the **Next** button to move forward to the next screen in the animation.

Click the **Back** button to replay any screen or to move back to the previous screen. Click on any **hyperlinks** to see an explanation of the word or phrase.

At the end, click **Replay** to watch the entire animation over again.

GRADEBOOK

From the **Syllabus**, click the **Gradebook** button at the top right to check your grades.

Use the scroll bar to see all the grades for your course. Click the **Refresh** button at the bottom to ensure that you are viewing your most current grades.

Click the **Syllabus** button to return to the syllabus screen.

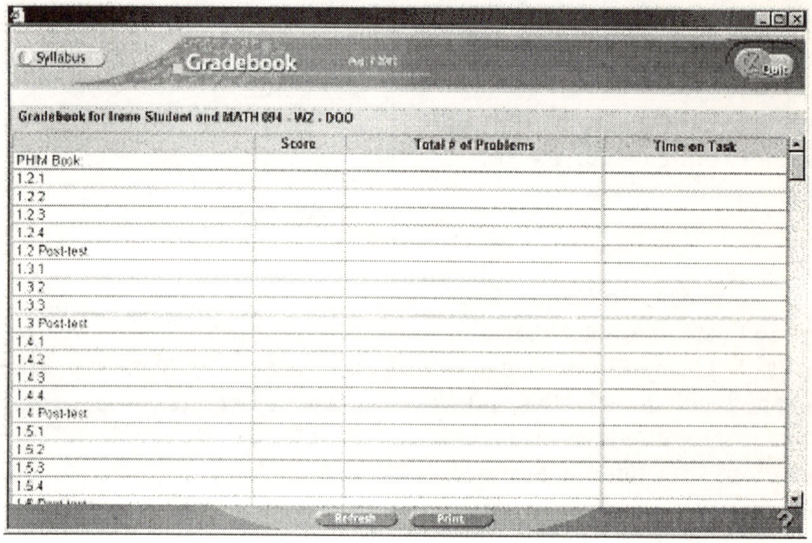

MESSAGES

From the Syllabus screen, click the **Messages** button at the top right to send a message or read a message sent by your instructor.

Click the **Return to Syllabus** button to go back to the Syllabus screen, or click the **Return to My Courses** button to go back to the My Courses screen.

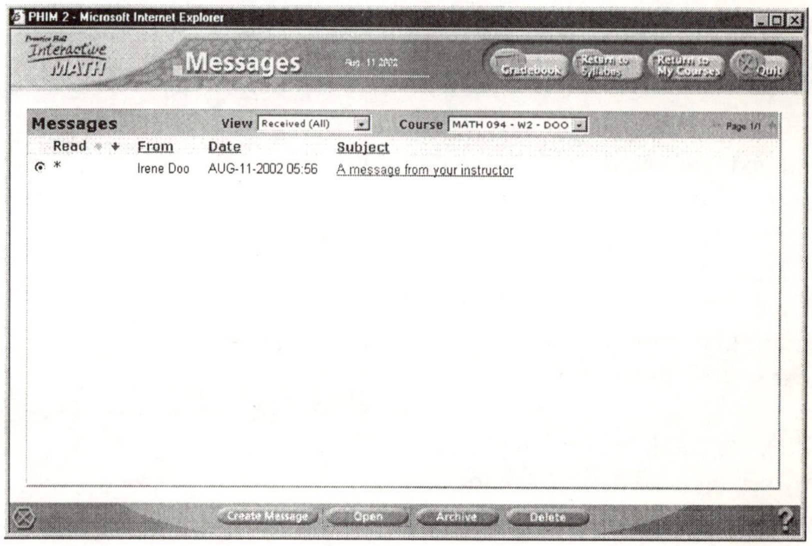

To read a message, select the radio button to the left of the message and click the **Open** button at the bottom of the screen. A pop-up window opens with the message, as shown on the right.

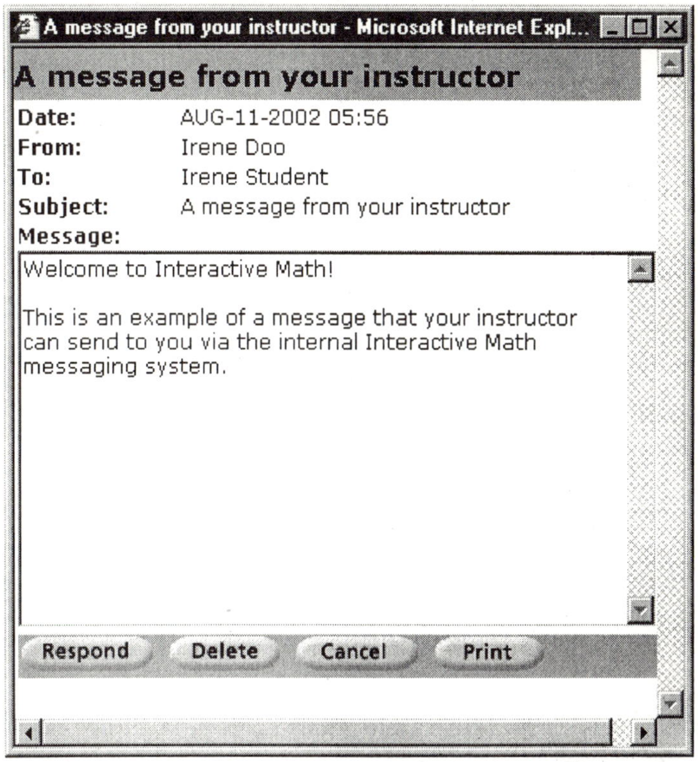

Click the **Respond** button to reply to the message.

Click the **Delete** button to delete the message.

Click the **Print** button to print a copy of the message.

Click the **Cancel** button to close the window and return to the **Messages** screen.

Click the "**Create Messages**" button at the bottom of the Message screen to send a new message to your instructor.

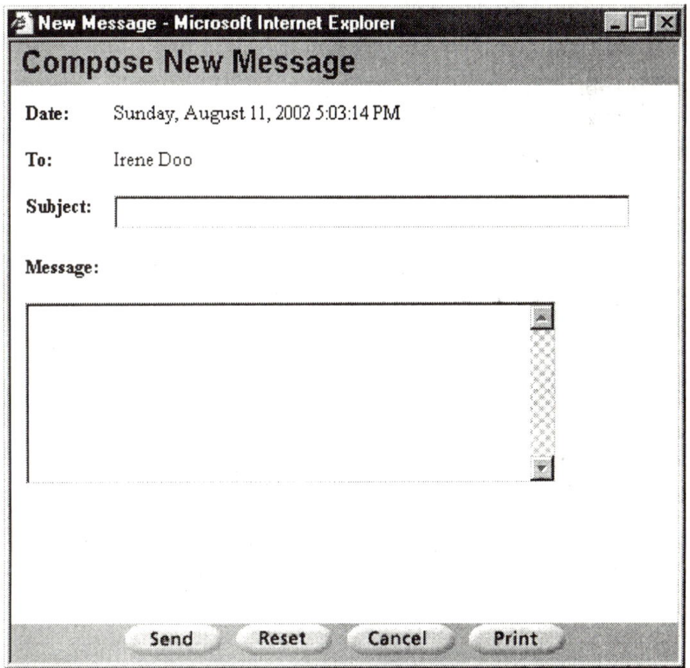

A pop-up window opens, as shown on the right. Type your subject and message, then click the **Send** button to send the message to your instructor.

Click **Reset** to erase the subject or message and start over.

After you have read a message, you can choose to store it in a different location. Click the **Archive** button to save the message in the Archive area for retrieval at any time.

You can also choose to delete a message by selecting it and clicking **Delete**.

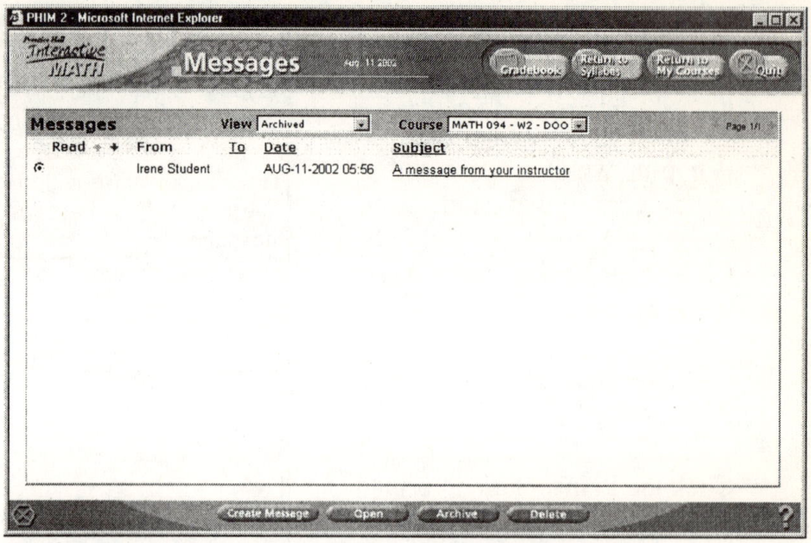

MATH HELP REQUEST

To send a help request to your instructor, click the **Math Help Request** button at the bottom of the screen. This option appears on the Practice, Assess, and Step by Step screens. The **Help Request** pop-up window opens up, as shown on the right.

Type your message and click the **Send** button. Be sure to include the Problem Number. Your instructor will receive your message together with the screens you are working on.

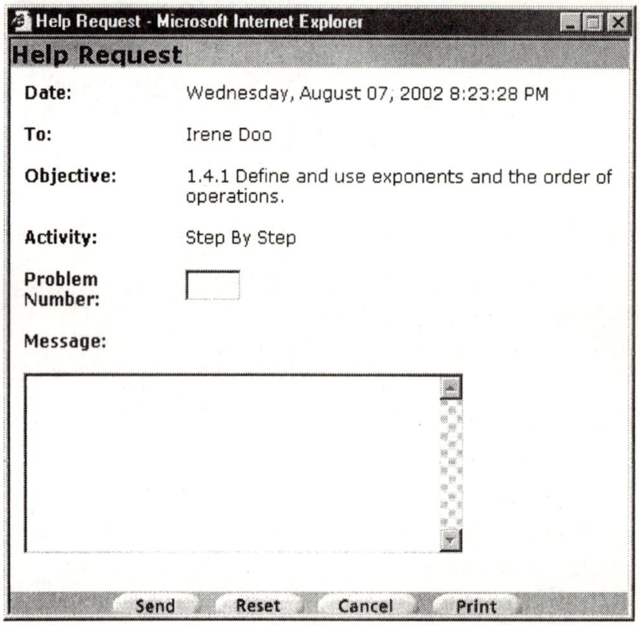

VOCABULARY

At any point within an objective, you can click the **Vocabulary** button to access definitions, formulas, and examples.

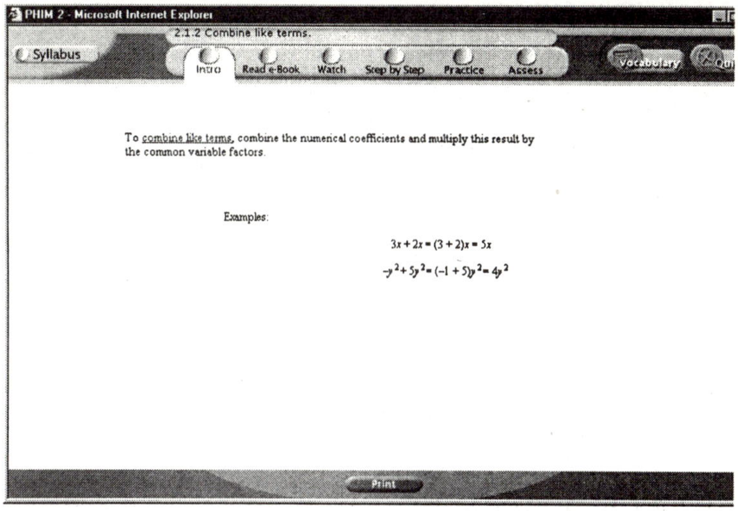

If you click on the **Vocabulary** button, you will see the screen immediately to your right.

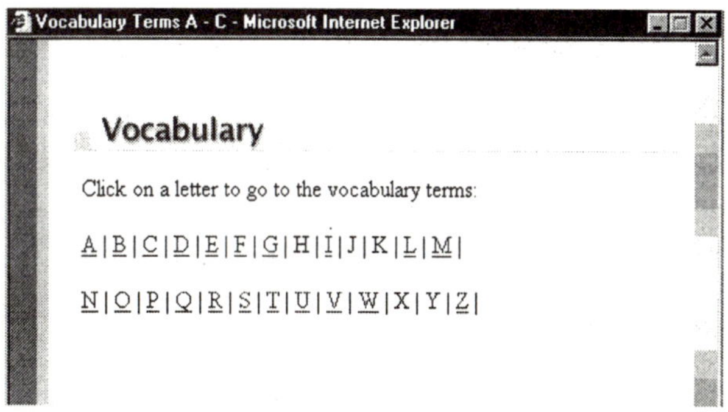

Click on the letter of your choice to find the definition and/or example of the concept you need.

The screen for the letter "I" appears on your right.

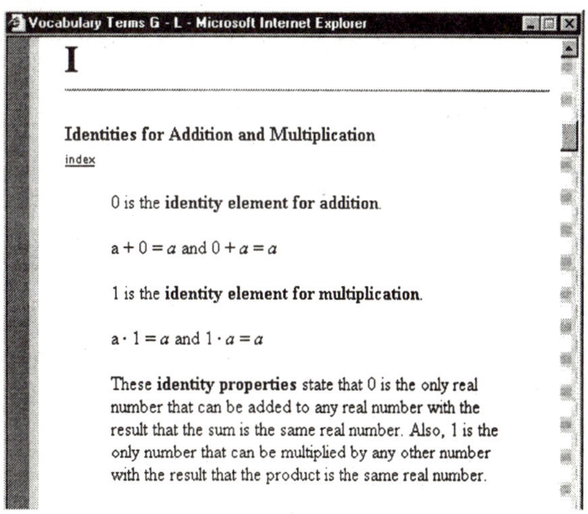

If you click on a blue hyperlinked term in a **Read e-book**, **Intro**, or **Demonstrate** screen, you will be sent to a screen such as this for a definition of the concept.

SYLLABUS ICONS

Closed, Opened, Completed Headings—A *closed book* icon indicates that there are content items located within this title that are not currently being displayed. An *open book* is the fully expanded list of items within that title and all are displayed. A *closed book with a blue check* indicates that all items below this level have been completed.

Incomplete, Complete, and Mastered Items—A *blank sheet of paper* icon indicates that this is a content item that has not yet been completed or mastered. A *paper with a blue check* icon is a completed content item. A *paper with a green check* icon indicates that the student has demonstrated mastery of this content item after taking a diagnostic test.

Syllabus Status Icons—A *question mark* icon indicates that a Help Request has been sent on this content item. A *paper with an "A"* icon shows that a response has been sent back to the student on a particular Help Request. A *clock with a red exclamation point* icon indicates that the target completion date for this item has passed, and it is not complete.

Chapter Opener and Lecture Video Icons—
The green and white block icon shows that you have a chapter opening animation included in your syllabus. The video camera icon indicates that you have a lecture video for a particular section.

Instructor Notes—A *blue circle with an "I"* button indicates that there is an Instructor Note for this content item.

PROGRAM HELP

Any time that you need technical or user help on the Prentice Hall Interactive Math 2 program, you can click on the "?" icon located at the bottom right of all content screens.

YOU CAN DO IT!—A CHECKLIST FOR SUCCESS

☑ **Follow the Syllabus.**

This is your course Syllabus, your timeline, your list of assignments, and your structure for the entire course. Make the Syllabus a frequent stop within the Prentice Hall Interactive Math 2 course.

☑ **Communicate with your instructor.**

Send Math Help Requests to your instructor if you need help solving a problem. Make sure that you follow your instructor's advice in any Instructor Notes on a particular content item within your Syllabus.

☑ **Use the resources available to you.**

Whenever you need to clarify a concept, review a definition or formula, or search for technical help navigating through the software, be sure to use the Program Help or "?" button and the Vocabulary button. The print supplements also contain additional reference resources such as appendices, extra problems, and solutions to problems. Take advantage of the support materials available to you.

☑ **Use the Demonstrate and Step by Step features.**

Use these features to further your understanding with extra guidance and examples of different problem-solving strategies.

☑ **Practice what you learn.**

Take advantage of the Practice and Assess options in every objective and try to do these exercises to test your knowledge of a particular concept. Use the exercises in the print supplements for additional practice.

☑ **Try all of the available options in each chapter.**

Use the different options that are available to you. Study the Intro, Read e-book, Watch, Step by Step, Demonstrate, Practice, and Assess screens in your objectives. Use the print supplements to enhance your work in the course. You have many resources available to you—all of them are designed to help you succeed.

☑ Know which learning style works best for you and use it.

Do you learn better by hearing things? Then listen as you work through the Demonstrate feature. Do you learn better by jumping in and exploring? Then use the Practice and Assess problems to try out ideas. If you are a visual learner, then any aspect of Prentice Hall Interactive Math 2 will appeal to you since visualization is a key part of the entire program.

☑ Manage your time.

The Syllabus only outlines your course by content item. Make sure that you try to keep the deadlines that your instructor sets for you. You will be more successful in your course if you learn to manage your time.

☑ Share your ideas.

Other students can be great resources for extra help, review, and hints on navigating through the software, working on group activities in the print supplements, or just explaining concepts. Ask your instructor about ways that you can work with other students to achieve your goals.

Keep this checklist in a place where you can read it frequently, and remember that you can succeed using Prentice Hall Interactive Math 2!

ISBN 0-13-102018-8
90000
9 780131 020184